Advanced Placement Examination in Chemistry

Frederick J. Rowe
Northport High School
Northport, NY

With the assistance of:
Sherry Berman-Robinson
Carl Sandburg High School
Orland Park, IL

ARCO

New York

First Edition

Copyright © 1990 by Frederick J. Rowe
All rights reserved
including the right of reproduction
in whole or in part in any form

 ARCO

Simon & Schuster, Inc.
15 Columbus Circle
New York, NY 10023

DISTRIBUTED BY PRENTICE HALL TRADE SALES

Manufactured in the United States of America

1 2 3 4 5 6 7 8 9 10

Library of Congress Number 89-489177
ISBN 0-13-010448-5

DEDICATED TO the more than 18,000 students and the AP chemistry teachers who toil yearly to master the college level chemistry course which is called AP Chemistry. Take the challenge of a foreign language, add a good dose of required mathematical skill and top it all off with many abstract concepts and you have the syllabus.

Acknowledgements

An AP Chemistry review book has been needed for some time. Publishers have been justifiably reluctant to take on such a project since the mechanicals for chemistry books are expensive. It requires the extensive use of subscripts, superscripts, diagrams, equations and charts to clearly present the information.

But now there is desktop publishing. This book was written, edited, designed, illustrated, and typeset on a Macintosh® computer. The mechanicals were produced on a laser printer.

I wish to thank Linda Bernbach, Senior Editor of ARCO books for recognizing the potential of the project. She enthusiastically guided the idea to a successful approval by the publisher and made sure that we produced a book that is as accurate and attractive as possible.

Sherry Berman-Robinson of Carl Sandburg High School (Orland Park, IL) has a special talent for preparing students successfully for the AP examination, and especially for the Part C Equations. She has shared some of her secrets by writing the draft for the chapter in Part III.

Bob Sims of the Westminster School (Atlanta, GA) has extensive expertise as a chemistry teacher and as the chair of the ACS Examinations Institute Subcommittee II (Advanced). He has contributed to this review by carefully critiquing the review chapters.

Helen Stone of Ben L. Smith High School (Greensboro, NC) spent many hours carefully reviewing the questions and must have worn out many red pens. Her many suggestions have been incorporated into the multiple choice questions, problems and essays.

And a special thanks to my colleague, friend and wife Doris (who teaches AP Studio in Art at Northport High School). Her support and philosophy has been, as always, indispensable.

Contents

Acknowledgements ix

Introduction xi

Part I **Review of Selected Topics**

Chapter 1 **Atomic Structure and Periodicity**
1. Atomic Theory and Atomic Structure 1
2. Quantum Numbers and Atomic Orbitals 2
3. Determining Electron Structure 3
4. Ion Structure 5
5. Relationships in the Periodic Table 5
6. Nuclear Chemistry 7

Chapter 2 **Chemical Bond**
1. Chemical Bonding 8
2. Octet Formulas and Lewis Dots 9
3. Expect the Unexpected 11
    Multiple Lewis Structures of the Same Compound 12
    Resonance 12
    Odd–electron Molecules 12
    Molecules with Incomplete Octets 13
    Central Atoms with More Than An Octet 13
4. Shapes of Molecules and Ions 14
5. Molecular Orbitals 17
6. Hybrids, Sigma Bonding and Pi Bonding 19
7. Coordination Chemistry 20

Chapter 3 **Stoichiometry**
1. The Mole Concept and Molar Mass 23
2. Empirical and Molecular Formulas 24
3. Reaction Stoichiometry 26
4. Limiting Reactants 27
5. Molarity, Molality and Mole Fraction 28

	6. Solution Stoichiometry	30
	7. Colligative Properties	30

Chapter 4   **States of Matter**

1. The Solid State — 32
2. Changes of State — 33
3. The Kinetic Molecular Theory of Gases — 34
4. The Ideal Gas Law — 35
5. Dalton's Law of Partial Pressures — 36
6. Properties of Solutions; Raoult's Law — 36

Chapter 5   **Reaction Kinetics**

1. Rate Equations — 38
2. First Order Reactions — 40
3. Activation Energy and Catalysis — 41
4. Heat Effects and the Arrhenius Equation — 43
5. Reaction Mechanism — 45

Chapter 6   **Equilibrium**

1. Equilibrium Constants — 47
2. Concentration Changes in Reactions — 48
3. LeChatelier's Principle — 50
4. Temperature Change and the Arrhenius Equation — 51
5. Solubility Equilibria and K_{sp} — 52
6. Common Ion Effect — 53
7. Selective Precipitation — 54
8. Mixtures of Two Solutions and "Bounce Back" — 55
9. Strong Acids and Strong Bases — 57
10. Weak Acids, Weak Bases and pH — 59
11. Buffers — 62
12. Titration and Neutralization — 63

Chapter 7 **Thermodynamics**

    1. Calorimetry, Internal Energy and Enthalpy — 65

    2. Hess' Law and Heat of Reaction — 67

    3. Heats of Formation, Combustion and Reaction — 69

    4. Bond Energy — 70

    5. Entropy — 71

    6. Free Energy — 72

    7. Free Energy and Equilibrium — 73

    8. Free Energy Change and Net Cell Potential — 75

Chapter 8 **Electrochemistry**

    1. Balancing Oxidation-Reduction Reactions — 76

    2. The Faraday and Electrolytic Cells — 79

    3. Voltaic Cells — 83

    4. Standard Reduction Potentials — 85

    5. The Nernst Equation — 87

Part II **Multiple Choice Questions**

    About Section I: The Multiple Choice Questions — 91

Set 1 **Atomic Structure and Periodicity** — 92
    Answer Key and Explanations — 96

Set 2 **Chemical Bond** — 98
    Answer Key and Explanations — 101

Set 3 **Stoichiometry** — 103
    Answer Key and Explanations — 106

Set 4 **States of Matter** — 109
    Answer Key and Explanations — 113

Set 5	Reaction Kinetics	115
	Answer Key and Explanations	120
Set 6	Equilibrium	122
	Answer Key and Explanations	126
Set 7	Thermodynamics	129
	Answer Key and Explanations	134
Set 8	Electrochemistry	137
	Answer Key and Explanations	141

Part III Free Response Section

About Section II: Free Response Problems — 147

Chapter 1. Part A Sample Problems
1. Gaseous Equilibrium — 149
2. Solubility Equilibrium — 149
3. Acid-Base Equilibrium — 150
   Sample Answers and Explanations — 150

Chapter 2. Part B Sample Problems
1. Electrochemistry–Electrolytic Cells — 154
2. Electrochemistry–Voltaic Cells — 154
3. Kinetics — 155
4. Stiochiometry—Empirical Formula and Colligative Properties — 155
5. Stoichiometry—Standard Solutions — 156
6. Thermodynamics — 156
   Sample Answers and Explanations — 157

Chapter 3.	Part C: Equations	
	1. General Considerations.	161
	2. Solubility Rules for Salts in Water Solutions.	162
	3. Reactions Involving No Changes in Oxidation States.	163
	4. Redox Reactions (Changes in Oxidation States)	165
	5. Practice Equation Writing	167
	Sample Answers and Explanations	169
Chapter 4.	Part D: Essays	174
	Sample Essay Questions	176
	Sample Answers and Explanations	181

Part IV   **Appendix**   187

Introduction

The Advanced Chemistry Review book is written in response to a need to have in one place materials that students will use to prepare efficiently for the examination. A careful investigation of the concepts which should be mastered is the focus of the topics covered in the review. A distillation of current college chemistry courses cannot replace the more complete coverage of topics that is possible during a year's study using a 700-page textbook.

It is neither required, nor expected, that candidates will have mastered the concepts covered in all of the texts. Students awarded the highest score on the AP test do not answer every question correctly. In fact, candidates who answer a little more than half of the Section I multiple choice questions correctly—depending on an equivalent success with the Section II free response questions—can anticipate their effort will be rewarded with a score of 5.0.

The book consists of three parts. The questions and problems are modeled on the AP style, but are not questions which have been used in previous examinations.

Part 1: Review of selected topics.

Part 2: Multiple–choice questions:
Strategy, questions, answers and explanations.

Part 3: Free response section II:
Questions, answers and explanations

Practice questions in addition to those supplied in this book are easy to get.

The College Board has published the:

"Entire 1984 AP Chemistry Examination and Key"
(Item 468223.....$5.00).
They soon will release another, more recent, examination.

Also available from the College Board is a:

"Set of free–response questions used in recent years"
(Item 254932.....$2.00).

Write to the:

Advanced Placement Program
CN 6670
Princeton, NY 08541-6670

The American Chemical Society prepares and publishes standardized multiple-choice tests. The proper level for AP preparation are the tests published biennially by the ACS Examinations Institute and entitled "High School Subcommittee II (Advanced)". The exams are confidential, so a teacher or school official must order and supervise the use of the more recent 1986ADV, 1988ADV or 1990ADV examinations. Earlier examinations are not as confidential.

Some students contend that the ACS/NSTA ADV questions are more rigorous than those used in Section I of the AP examination.

Write to the:

ACS DivCHED Examinations Institute
107 Physical Sciences
Oklahoma State University
Stillwater, OK 74078

The explanations in this book use problem-solving techniques that can be applied to questions dealing with several topics. Current examination usage of units is maintained. A unit like molarity, M, is written in the exam as moles L^{-1}. This should be read as moles per Liter.

The AP examination is a race against the clock. It has been assumed that the candidate who purchases this book will, like an athlete, make sure to have the right equipment. This should include bringing to the exam, and knowing how to use, a hand-held, non-programmable scientific calculator. The calculator should be capable of (in addition to performing the usual addition, multiplication, etc.) computing exponential numbers, logarithms (both ln x and log), powers and roots. No provision has been made in this review to explain how to 'do-it-by-hand' or to explain the use of log-tables.

Part I

Review of Selected Topics

Chapter 1
Atomic Structure and Periodicity

1. Atomic Theory and Atomic Structure

An atom is electrically neutral since it contains the same number of protons and electrons. The number of protons (the atomic number) determines the identity of the element. The mass number is the sum of the number of protons and neutrons.

Example

Element	Protons	Electrons	Neutrons	Mass #
In	49	49	71	120
Sn	50	50	70	120
Sb	51	51	70	121
Sb	51	51	69	120
Te	52	52	68	120

Atoms of a particular element may vary in mass (isotopes) because of a differing number of neutrons in the nucleus. Particular isotopes are identified by the symbol of the element with the atomic number at the lower-left corner and the mass number at the upper left corner.

Example

Element	Protons	Neutrons	Mass#	Symbol
Sn	50	69	119	$^{119}_{50}Sn$
Sn	50	70	120	$^{120}_{50}Sn$
Sb	51	70	121	$^{121}_{51}Sb$
Sb	51	69	120	$^{120}_{51}Sb$
Te	52	68	120	$^{120}_{52}Te$
Te	52	72	124	$^{124}_{52}Te$

Review of Selected Topics

The weighted-average of the naturally occurring isotopes of an element determines the atomic weight.

Example

Element	Mass #	Abundance	Mass x Abund.
Cu	62.9296	69.20%	4355
Cu	64.927	30.80%	2000

Weighted Average = Sum ÷ 100	63.54

Atoms may gain or lose electrons to form electrically charged species called 'ions'.

Example

Element	Protons	Electrons	Mass#	Symbol
Na	11	10	23	$^{23}_{11}Na^{+1}$
Ga	31	28	70	$^{70}_{31}Ga^{+3}$
S	16	18	35	$^{35}_{16}S^{-2}$
Br	35	36	81	$^{81}_{35}Br^{-1}$
Pb	82	78	206	$^{206}_{82}Pb^{+4}$

2. Quantum Numbers and Atomic Orbitals

Electrons in atoms are quantized, which means that electrons in atoms have only certain allowed energy states. The energy of an electron in an atom is determined mainly by the value of n the principal quantum number. The principal quantum number defines a shell, at some average distance from the nucleus, capable of holding electrons having about the same energy. The value of the principal quantum number can be any positive integer. The lowest energy electrons have n=1. Only values up to n=7 are necessary to describe the electrons in the first 118 elements.

The shells consist of one or more sub-shells determined by the orbital quantum number, l. Allowed values of l include any positive integer from zero (0) to (n-1). Usually, instead of integers, values of l are assigned the letters s, p, d and f. Orbital quantum number (l) values of 0, 1, 2, and 3 are all that are necessary to list the sub-levels of the known or soon to be known first 118 elements.

Not only does the orbital quantum number indicate differences in energy of electrons within a shell, but it also indicates the shape of the orbital.

l	0	1	2	3
letter	s	p	d	f
shape	spherical	dumbells	N.A.	N.A.

Note: N.A. means not appropriate for testing at the A.P. level.

Each sub–shell consists of one or more orbitals of the same energy indicated by the magnetic quantum number, m_l. Permitted values of m_l are integers from $-l$ to $+l$.

Summary for orbitals of the first four periods.

Shell n	Subshell l	Orbital Letter	m_l	No. of Orbitals
1	0	1s	0	1
2	0	2s	0	1
	1	2p	-1 0 +1	3
3	0	3s	0	1
	1	3p	-1 0 +1	3
	2	3d	-2 -1 0 +1 +2	5
4	0	4s	0	1
	1	4p	-1 0 +1	3
	2	4d	-2 -1 0 +1 +2	5
	3	4f	-3 -2 -1 0 +1 +2 +3	7

Electrons behave as if they were spinning. Two electrons can occupy the same orbital, but they must spin in opposite directions, The electron spin is indicated by the spin quantum number, s, which has values of $\pm\frac{1}{2}$.

3. **Determining Electron Structure**

The electronic configuration of any atom in its ground (most stable and lowest energy) state can be determined by using the aufbau (building up) procedure. Two electrons are added per orbital (lowest energy orbitals first) until the number of added electrons equals the number of protons in the atom (making the atoms electrically neutral).

Electrons placed in a subshell consisting of more than one orbital will go into a empty orbital rather than pair up with the first electron in a half–filled orbital (Hund's Rule). And its spin direction will be the same as that of the first electron (parallel spin). Electron configurations and the order of filling subshells can readily be obtained from a periodic table.

Review of Selected Topics

Note: n is the period number

Block	Groups	Configuration
s-block:	IA through IIA	ns^1 through ns^2
p-block:	IIIA through VIIA plus Group 0 (Noble Gases)	$ns^2 np^1$ through $ns^2 np^6$
d-block:	Transition elements IIIB through IIB	***$(n-1)d^1 ns^2$ through ***$(n-1)d^{10} ns^2$
f-block:	Inner Transition elements Lanthanide and Actinide series	$(n-2)f^1 (n-1)d^1 ns^2$ through $(n-2)f^{14} (n-1)d^1 ns^2$

***Some irregularity exists, particularly in Groups IB and VIB.

The periodic table shows how the blocks are positioned.

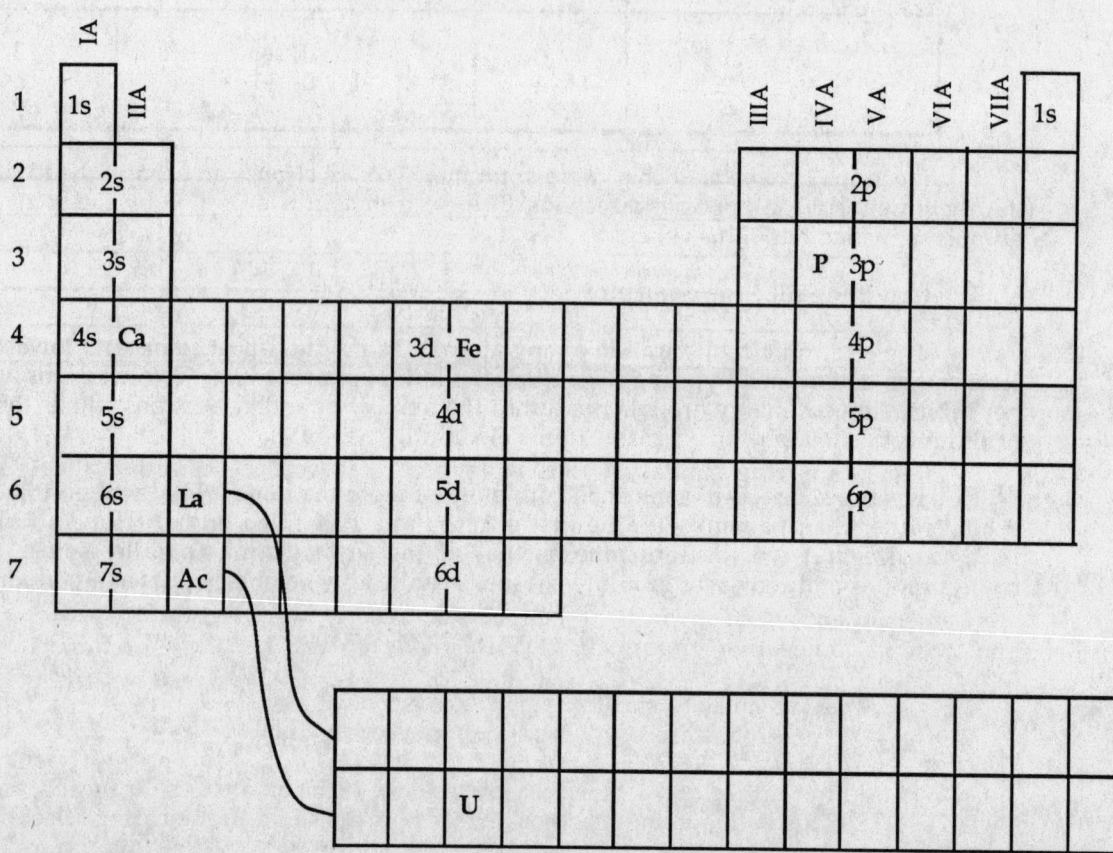

Example

Use the periodic table above to determine the outer electron structure for calcium (Ca), iron (Fe), phosphorus (P) and uranium (U).

Calcium is [Ar]$4s^2$. Elemental calcium found in the second box of the s-block of period 4.

Iron is [Ar]$3d^6\ 4s^2$. Iron is located in the sixth box of the *d*-block of period 4, after passing through the $4s$-block. The outer, $4s$-electrons are written last, even though they are filled before the $3d$-electrons

Phosphorus is [Ne]$3s^2\ 3p^3$. The element phosphorus is found in the third box of the *p*-block of period 3, after passing through the $3s$-block.

Uranium is [Rn]$5f^3\ 6d^1\ 7s^2$. Uranium is located in the third box of the *f*-block of period 7, after passing through the $7s$-block and through the first box of the $6d$-block.

4. Ion Structure

Electron configurations for ions can be predicted from their respective atoms. Positive ions are formed when highest energy (generally the outermost) electrons are lost. Negative ions are formed when electrons are added to the outermost subshell.

Examples

Element	Structure	Ion	Structure
Ca	[Ar]$4s^2$	Ca^{2+}	[Ar] or [Ne]$3s^23p^6$
S	[Ne]$3s^23p^4$	S^{2-}	[Ne]$3s^23p^6$ or [Ar]
Mn	[Ar]$3d^54s^2$	Mn^{2+}	[Ar]$3d^5$
Ti	[Ar]$3d^24s^2$	Ti^{3+}	[Ar]$3d^1$
Ag	[Kr]$4d^{10}5s^1$	Ag^{1+}	[Kr]$4d^{10}$

5. Relationships in the Periodic Table

Elements having similar properties are placed in vertical columns called 'Groups'. Horizontal rows, which start when there is a repeat of properties, are called 'Periods'.

The <u>radii of atoms and ions</u> decrease as atomic numbers increase from left to right across a period. The cause is an increased attraction between the increased number of protons in the nucleus and electrons in the valence shell. There is a discontinuity where a change from positive to negative ion formation occurs (at Group VA).

1. Positive ions are smaller than their respective atom.
2. Negative ions are larger than their respective atom.

The radii of atoms and ions increase as atomic numbers increase from top to bottom down a group. This is due to an increased number of shells and the shielding effect of all the inner shells.

Ionization energy is a measure of the tightness with which electrons are held by an atom or ion. The ionization energy of atoms and ions increases as the atomic number increases from left to right across a period. There is a discontinuity when there is a change from positive to negative ion formation in the middle of the table.

1. Electrons are held more tightly by positive ions.
2. Electrons are held less tightly by negative ions.

The ionization energy of atoms and ions decreases as the atomic number increases from top to bottom down a group.

Electron affinity is the measure of the attraction of electrons by an atom. The electron affinity of atoms follows the same trend as the ionization energy.

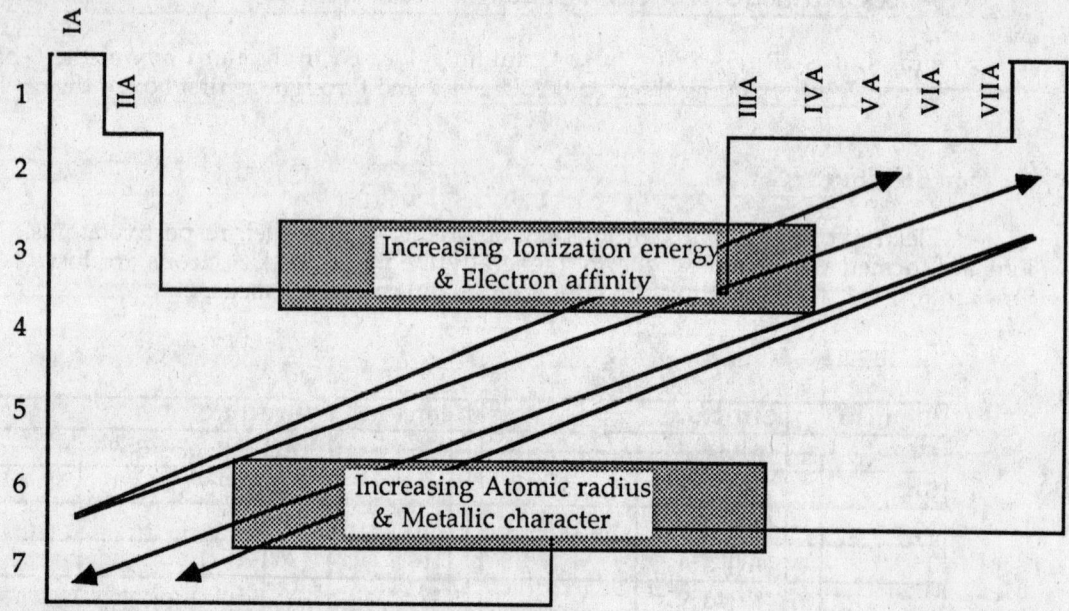

Example

Arrange sets of three elements in order of increasing size and increasing ionization energy (and electron affinity).
- a. Be, Mg, Ca
- b. S, Cl, F
- c. As, N, F
- d. Ga, Ge, In
- e. Te, I, Xe

Increasing Size	Increasing Ionization Energy
Be < Mg < Ca	Ca < Mg < Be
F < Cl < S	S < Cl < F
F < N < As	As < N < F
Ge < Ga < In	In < Ga < Ge
Xe < I < Te	Te < I < Xe

6. Nuclear Chemistry

When a natural radioactive atom decays, its nucleus loses either an alpha or a beta particle. If the nucleus of an atom is bombarded by an alpha particle, a beta particle, a neutron, a proton or a deuteron, the bombarding particle may be captured by the nucleus, forming a new element.

Particle	Symbol
alpha	$^{4}_{2}He$
beta	$^{-1}_{0}e$
neutron	$^{1}_{0}n$
proton	$^{1}_{1}H$
deuteron	$^{2}_{1}H$

Nuclear equations are written by recognizing that the total of the atomic numbers values and the mass number values must be the same on both sides of the equation.

Example

$$^{14}_{7}N + {}^{4}_{2}He \longrightarrow {}^{18}_{9}F \longrightarrow {}^{17}_{8}O + {}^{1}_{1}H$$

The rate of radioactive decay is expressed in terms of half-life. The same equations are used for first order reactions that are studied in both chemical kinetics and radioactive decays.

$$\ln \frac{N_t}{N_o} = -\lambda t \qquad \lambda = \frac{0.693}{t_{1/2}} \qquad \ln \frac{N_t}{N_o} = -\frac{0.693\,t}{t_{1/2}}$$

N_t is the number of moles at time t.
N_o is the original number of moles.
N_t/N_o is the mole fraction remaining at time t.
λ is the decay constant and $t_{1/2}$ is the half-life.

Example

Radioactive sodium-24 has a half-life of 15.0 hours. How many days will it take for the radioactivity to fall to 0.10 of its original activity?

$$\ln 0.10 = - \frac{0.693\,t}{15.0 \text{ hrs}}$$

Answer: time (t) = 50. hrs = 2.1 days

Chapter 2
Chemical Bond

1. Chemical Bonding

Our current model of the chemical bond makes use of the definitions of the three strong bonds between atoms or ions; the metallic, the ionic and the covalent bond.

Metallic bonding

The bonding between metals must account for the typical physical properties associated with metals:

.....uniform, high electrical and heat conductivity

.....high malleability and ductility.

.....shininess and luster.

The simpler 'electron–sea' definition is still useful, and describes the metallic bond as being cations or positive ions (the 'kernel') surrounded by a 'sea of mobile electrons'. Mobile electrons will explain the easy movement of heat and electricity. The 'kernel' can be easily realigned during compression or tension. The 'sea of mobile electrons' reflects light in many directions, causing metals to be shiny.

Ionic bonding

The formation of bonds between ions must account for the typical physical properties associated with salts:

.....their brittleness.

.....their lack of electrical conducivity in the solid state.

.....their solubility in polar solvents such as water.

.....their conductivity in aqueous solutions.

When a metal reacts with a nonmetal, the metal loses its valence electrons and the nonmetal gains these electrons. The valence shell of the nonmetal is completed and gives it a stable noble gas electron structure. The cations and anions formed are held together by electrostatic attraction. A strongly bonded crystalline structure of ions explains the brittleness and the lack of electrical conductivity in the solid state. The existence of ions accounts for the attraction to polar solvents. Dissolved ions are mobile charges and can conduct electricity.

Covalent bonding

Nonmetals have strong attractions for electrons as evidenced by their high ionization energy, electronegativity and electron affinity. A covalent bond is the sharing of a pair of electrons between two nonmetallic atoms. These electrons may be shared equally or unequally. Unequal sharing of electrons is predicted by an electronegativity difference and the result is a polar covalent bond with ionic character.

2. Octet Formulas and Lewis Dots

There are over two-million known molecular compounds, so it makes sense to know how to visualize and represent covalently bonded substances. Many molecules form electron-pair bonds such that each atom achieves a closed-shell electron configuration. Hydrogen needs two electrons to achieve a closed-shell and acquire the stable, noble-gas structure of helium. The other nonmetals, such as fluorine, require eight electrons (an 'octet') to achieve electron stability.

Hydrogen   H• + •H ⇒ H:H

Fluorine: :F• + •F: ⇒ :F:F:

In writing the Lewis structure for a compound, only include the outermost (or valence) electrons. For the atoms of the main-group elements the number of valence electrons is the same as the group number.

The covalent bond between atoms is ordinarily shown as a line between bonded atoms. The unshared electron-pairs (commonly called lone-pair electrons) are shown as dots. The example shows four common compounds containing a period 2 nonmetal and hydrogen. The central atom in each case is surrounded by an octet of electrons, either in pairs of shared electrons (a line) or in lone-pairs (two dots).

Example:

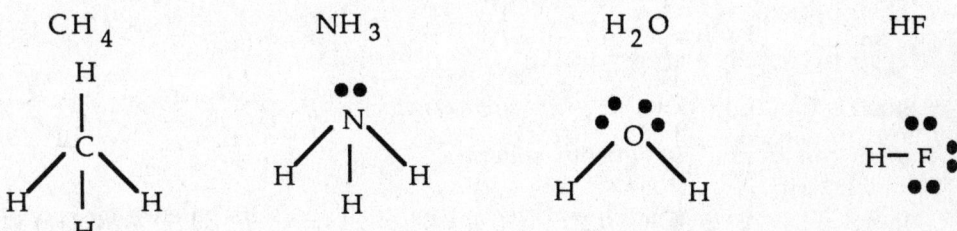

10 Review of Selected Topics

Lewis structures may be written by following the simple rules illustrated in the example below.

Example

1. Draw a structure for the molecule by joining atoms by single bonds at first.

    $CH_4 S$

    $$H-\underset{\underset{H}{|}}{\overset{\overset{H}{|}}{C}}-S-H$$

2. Count the number of valence electrons. Add electrons for negative ions.
   C (4) + 4H (4) + 1S (6) = 14

3. Deduct two electrons for each bond.
   14 - (5 * 2) = 4

4. Distribute the remaining electrons as lone–pairs to give each atom an 'octet' of electrons.

5. If there are too few electrons for 'octets', consider multiple–bonding.

    $$H-\underset{\underset{H}{|}}{\overset{\overset{H}{|}}{C}}-\overset{..}{\underset{..}{S}}-H$$

Example: Structural formulas with single bonds.

Formula	$POCl_3$	ClO_2^{1-}
Step 1	$Cl-\underset{\underset{Cl}{|}}{\overset{\overset{Cl}{|}}{P}}-O$	$[O-Cl-O]^{1-}$
Step 2	1P (5) + 3Cl (21) + 1O (6) = 32	1Cl (7) + 2 O (12) + 1 = 20
Step 3	32 - (4 * 2) = 24	20 - (2 * 2) = 16
Step 4	(Lewis structure of $POCl_3$ with lone pairs on all Cl and O)	(Lewis structure of ClO_2^{1-} with lone pairs)

Chemical Bond

Example: Structural formulas with multiple bonds

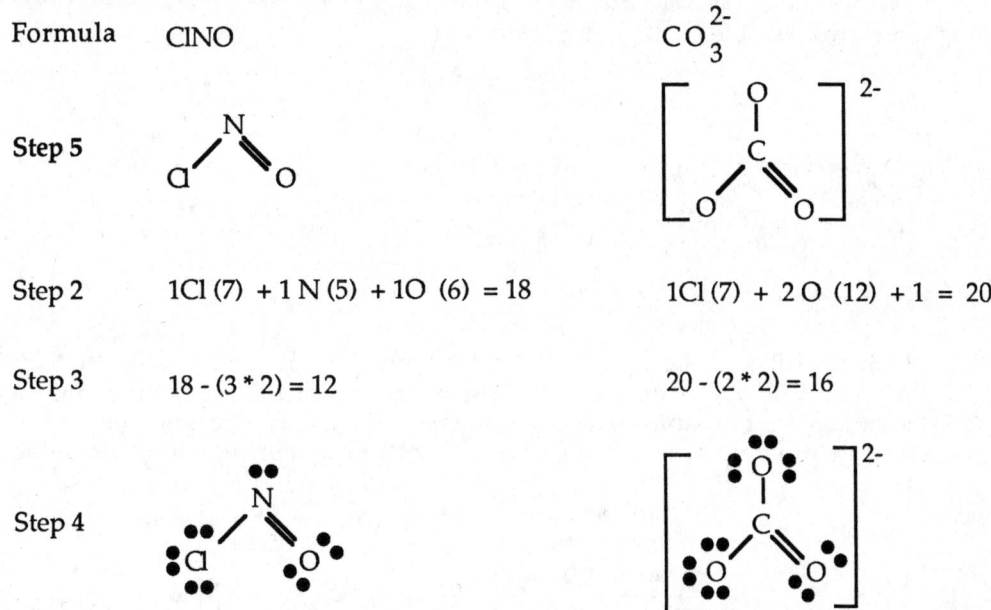

Formula   ClNO CO_3^{2-}

Step 5

Step 2   1 Cl (7) + 1 N (5) + 1 O (6) = 18   1 Cl (7) + 2 O (12) + 1 = 20

Step 3   18 - (3 * 2) = 12   20 - (2 * 2) = 16

Step 4

Lewis structures can correctly show the difference in bonding between an ionic compound and a covalent compound. The example shows this difference in each pair. (The first is the covalent compound.)

Examples:
1.   2.

HCl and NaCl SCl_2 and $CaCl_2$

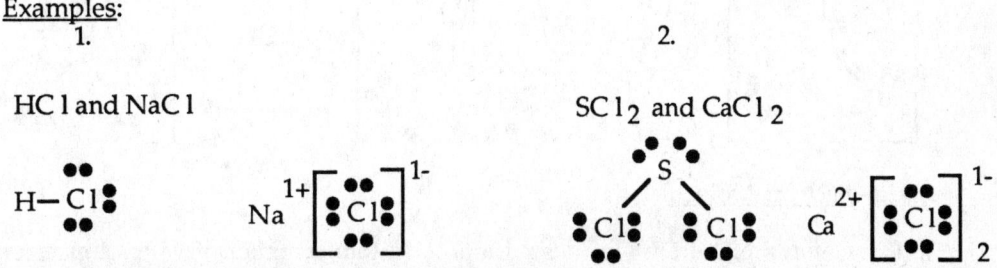

3.   **Expect the Unexpected (Lewis Formulas and The Octet Rule)**

The rules for Lewis structures may be followed explicitly to allow the correct representation of most molecules and ions. A number of unexpected answers come up often enough to warrant further discussion. The 'answer' can only be resolved with additional information concerning the physical and chemical properties of the molecule being described.

Multiple Lewis Structures of the Same Compound.

Diazomethane has the structural formula H₂CNN. The diagrams show three Lewis structures possible for the compound.

Resonance

When the rules for Lewis structures are applied to SO₂, SO₃, NO₂⁻ or NO₃⁻, two or three equivalent structures result. The properties of these structures indicate that all the bonds are the same. The phenomenon of resonance occurs when more than one structure can be drawn, but where one structure is not enough to describe the species.

Odd-electron Molecules

NO₂, CN and NO are molecules with an odd total number of valence electrons. There will be at least one unpaired electron in the structure and at least one atom lacking an octet of electrons. These molecules with unpaired electrons will be 'paramagnetic', as contrasted with 'diamagnetic' molecules whose electrons are all paired.

Molecules with Incomplete Octets

Beryllium and boron are representative of elements which can form covalent bonds, but whose central atom only has two ore three electrons, respectively. BeCl$_2$ and BF$_3$ have structures with less than an octet of electrons.

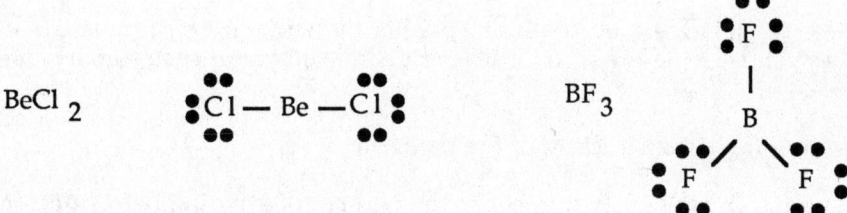

Central Atoms with More Than An Octet

Elements such as phosphorus (P) and sulfur (S) contain central atoms whose d-subshells are available at low enough energies to participate in bonding. In the case of PCl$_5$ the phosphorus has 10 electrons (5 bonds) surrounding it. The SF$_6$ has 12 electrons (6 bonds) around the sulfur.

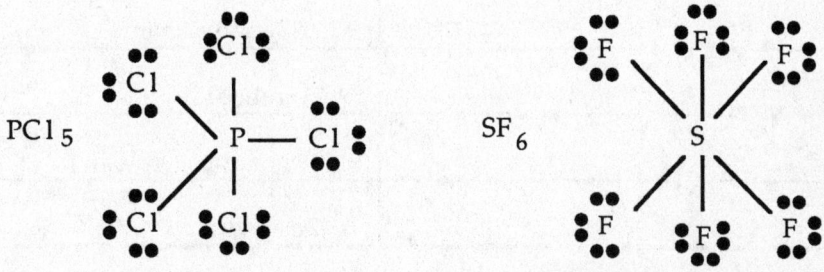

4. Shapes of Molecules and Ions: Valence Shell Electron-Pair Repulsion Theory (VSEPR)

The VSEPR model is a simple, effective way to predict the structure of molecules. These shapes play a very important role in predicting the chemical properties of nonmetallic compounds such as dipole moment. The main postulate of this model is that the bonding and nonbonding pairs around the cental atom will be positioned as far from each other as is possible.

Use the following procedure to predict the structure of a molecule using the VSEPR theory. As you read the rules, confirm your understanding of each step by consulting the table of examples.

1. Write the Lewis structure of the molecule.

2. Use the Lewis structure to count the number of bonding pairs and lone–pairs around the central atom.
A single, double, or triple covalent bond is counted as one bonding electron region. An unpaired electron is counted the same as a lone pair.

3. Identify the most stable arrangement of the electrons by counting the number of bonds.

Bonds and lone–pairs	Shape
2	Linear
3	Trigonal planar
4	Tetrahedral
5	Trigonal bipyramidal
6	Octahedral

4. If more than one arrangement of lone pairs and chemical bonds is possible, select the one that will minimize lone-pair repulsions.

   In trigonal bipyramidal arrangements, the repulsion is minimized when every lone pair is in the plane of the triangle.

   In octahedral arrangments the repulsion is minimized when the lone pairs are along the axis of the central atom.

5. The molecular structure is determined by the location of atoms at the end of chemical bonds. Lone–pairs are 'invisible'.

Example: The shape of molecules as a function of the number of bonding and nonbonding electron–pairs.

Number of electron pairs	Number of lone pairs	Shape of bonds	Example	Geometry	Bond Angle
2	0	linear	BeCl	Cl—Be—Cl	180°
3	0	trigonal planar	BF_3	F, B, F, F	120°
3	1	trigonal planar	SO_2	S, O, O (angular)	120°
4	0	tetrahedral	CCl_4	Cl, C, Cl, Cl, Cl	109.5°
4	1	tetrahedral	NH_3	N, H, H, H (trigonal pyramidal)	109.5°
4	2	tetrahedral	H_2O	H, O—H (bent or angular)	109.5°
4	3	tetrahedral	HF	H—F (linear)	180°

Review of Selected Topics

Number of electron pairs	Number of lone pairs	Shape of bonds	Example	Geometry	Bond Angle
5	0	trigonal bipyramidal	PCl_5	(trigonal bipyramidal with Cl atoms around P)	90° and 120°
5	1	trigonal bipyramidal	SF_4	irregular pyramid	90° and 120°
5	2	trigonal bipyramidal	ICl_3	T-shaped	90° and 120°
5	3	trigonal bipyramidal	I_3^-	linear	90° and 120°
6	0	Octahedral	SF_6	(octahedral)	90°
6	0	Octahedral	BrF_5	square pyramidal	90°
6	0	Octahedral	$XeF4$	square planar	90°

5. Molecular Orbitals

The molecular orbital (M.O.) model was developed to predict structure better than the resonance model. This theory handles molecules with unpaired electrons and with odd numbers of electrons better than the Lewis or VSEPR procedures. MO Theory allows correct estimates of the bond energy, the bond length and either paramagnetism (unpaired electrons) or diamagnetism (paired electrons) in molecules.

A set of molecular orbitals is formed from their corresponding atomic orbitals. These may be either bonding orbitals (between the atoms) or anti–bonding orbitals (either side of the atoms). A molecular orbital has energy, and the procedures used in understanding the electron structure of atoms are followed here also. The 'aufbau' principle, the Pauli Exclusion Principle and Hund's Rule are used in describing molecular orbital structures.

Use the following procedure to predict the structure the molecular orbital structure of a molecule. As you scan the examples, confirm your understanding of each step.

1. Count the electrons in the molecule or ion.

2. Electrons should be assigned to the lowest energy bonding ($\sigma_{2s}^b, \sigma_{2p}^b, \pi_{2p}^b$) or anti–bonding ($\sigma_{2s}^*, \sigma_{2p}^*, \pi_{2p}^*$) orbital available. Obey the Pauli Exclusion Principle (a maximum of 2 electrons per orbital) and Hund's Rule (electrons in equivalent energy orbitals occupy them singly first).

3. Bond 'order' = (no. of bonding electrons) − (no. of anti–bonding electrons)

4. Bonds = $\dfrac{\text{Bond order}}{2}$

5. Bond order is directly proportional to bond energy, and inversely proportional to bond distance. A bond order of 6 (a triple covalent bond) represents a molecule with a large bond energy and a small bond distance.

6. <u>Diamagnetism</u> exists when all the electrons in the molecule are paired.
   <u>Paramagnetism</u> exists when there are one or more unpaired electrons in the molecule.

Review of Selected Topics

<u>Examples:</u> The following structures exist for homonuclear species of the second period.

<u>Note:</u> There is a change in the order of filling that occurs when the structure of O_2 is written. Do not memorize the order of filling, but do be aware that it can change.

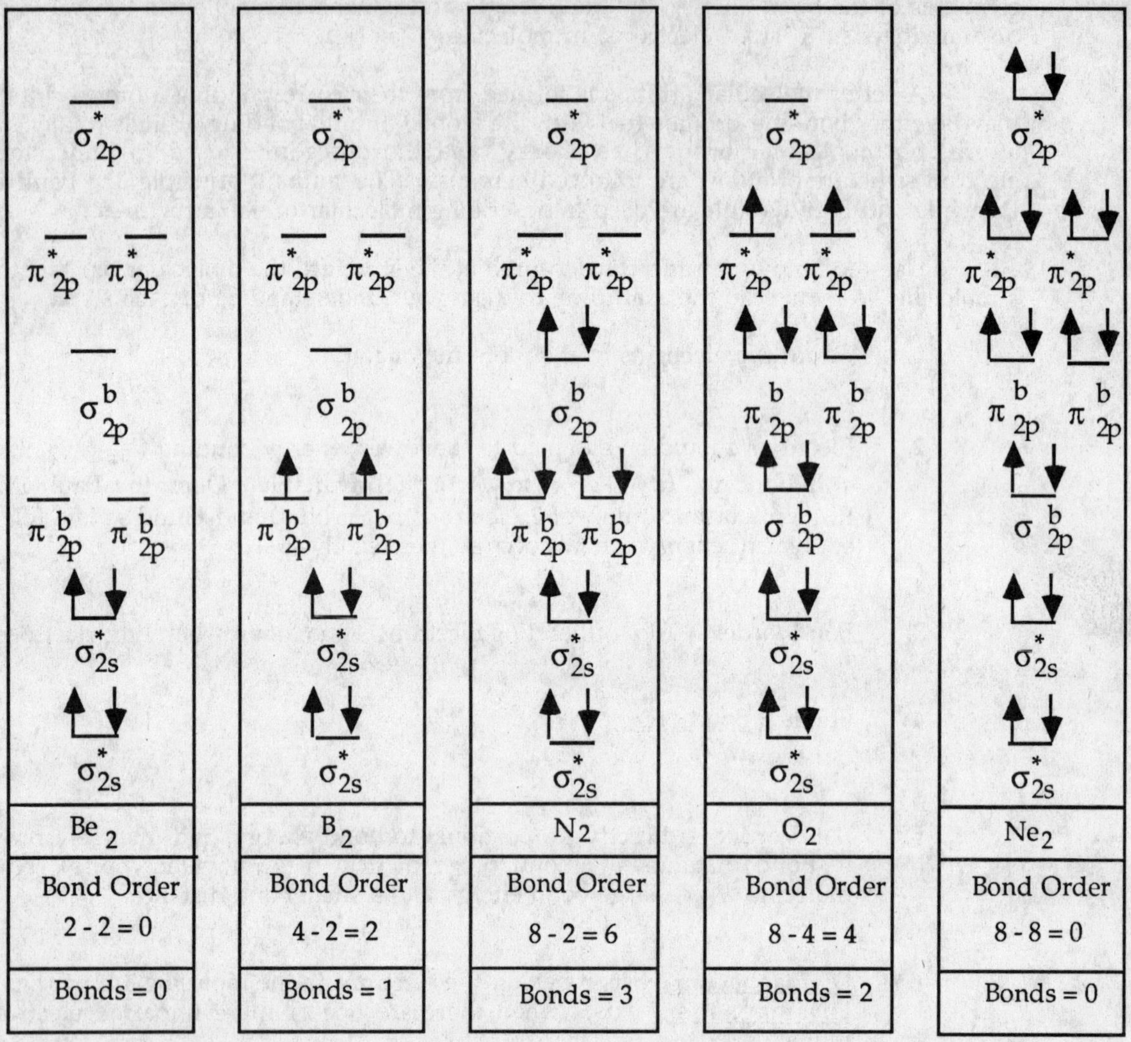

Be_2	B_2	N_2	O_2	Ne_2
Bond Order $2 - 2 = 0$	Bond Order $4 - 2 = 2$	Bond Order $8 - 2 = 6$	Bond Order $8 - 4 = 4$	Bond Order $8 - 8 = 0$
Bonds = 0	Bonds = 1	Bonds = 3	Bonds = 2	Bonds = 0

6. Hybrids, Sigma Bonding and Pi Bonding

A sigma (σ) bond occurs when two atomic orbitals combine to form a covalent bond that is along the axis connecting two atomic nuclei. It is the first bond that will be formed in the case of double or triple covalent bonds.

Subsequent bonds will be pi (π) bonds. A pi-bond is a sausage-shaped region above and below the nuclei of bonded atoms. A double covalent bond consists of one σ and one π bond, and a triple covalent bond consists of one σ and two π bonds.

An important effect of a π bonding orbital is to fix the σ bond in a plane, and to prevent rotation about a double-bond or a triple-bond.

A common occurrence during covalent bonding is the formation of sp, sp^2 or sp^3 'hybrid' atomic orbitals. To determine the kind of hybrid formed by the bonding of the carbon atom in a molecule, count the number of bonds. A double or triple covalent bond counts as <u>one</u> bond. Two bonds results in an sp, three bonds gives an sp^2 and four bonds produces an sp^3 hybrid.

In the table, the Lewis structure is first drawn for the molecule. The number of bonds is found by examining the bonding of each carbon atom, and the bonding hybrid of the carbon is determined by the number of bonds. The bond angle is set by the hybrid type. The number of sigma and pi bonds are then counted. The single bond between a hydrogen atom and a carbon atom counts as one sigma bond. The bonding between the carbon atoms accounts for the rest of the totals.

Structure	Number of bonds	Hybrid-ization	Bond Angle	Sigma bonds	Pi bonds
H₃C—CH₃ (ethane)	4	sp^3	109.5°	7	0
H₂C=CH₂ (ethene)	3	sp^2	120°	5	1
H—C≡C—H	2	sp	180°	3	2

Example:

Determine the electron hybridization for the central atom, the bond angles, and the number of sigma and pi bonds for CH_4, H_2CO, CO_2 and HCN.

Structure	Number of bonds	Hybrid-ization	Bond Angle	Sigma bonds	Pi bonds
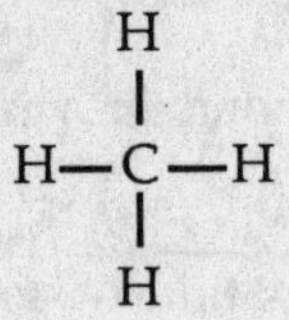 H—C(—H)(—H)(—H) with H above	4	sp^3	109.5°	4	0
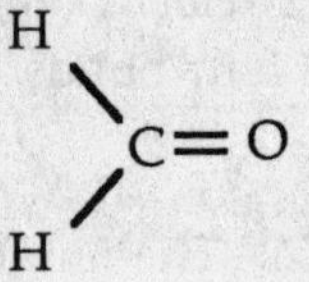 H₂C=O	3	sp^2	120°	3	1
O=C=O	2	sp	180°	2	2
H—C≡N	2	sp	180°	2	2

7. Coordination Chemistry

Transition metal complex ions are species in which ligands bond by contributing an electron-pair to the central atom or ion. The naming of these ions comes from an early model which generally assumed that coordinate covalent bonds are formed by involving the d-orbitals of the central metal ion. A complex ion is a distinct chemical, with properties different from its parts.

Many coordination compounds are chemically stable. The multiple oxidation states and electron structure of the transition metals often result in complexes which are vividly colored. The colors and stability make some complexes useful as confirmatory tests in schemes for determining the presence of ions in qualitative analysis.

Chemical Bond

Square planar and octahedral structures are the most common geometries for complex ions. Familiarity with the systemic naming of these species permits insight into their use in analytical procedures and a practical use for chemicals with an octahedral structural geometry.

Procedure for naming coordination compounds

1. The cation (- ion) is named before the anion.

2. The ligand is named before the metal.

3. End the name of a negative ligand with the letter 'o'. Give neutral ligands the name of the neutral molecule.

Common Monofunctional Ligands	
Ligand	Name
F^-, Cl^-, Br^-, I^-	Fluoro, chloro, bromo, iodo
$:NO_2^-$	Nitro
$:CN^-$	Cyano
$:OH^-$	Hydroxo
H_2O	Aquo
NH_3	Ammine

Common multifunctional ligands			
Symbol	Ligand	Formula	Bonds
en	Ethylenediamine	$NH_2-CH_2=CH_2-NH_2$	2
EDTA	Ethylenediamine-tetraaceto		6

4. Count the number of ligands present by using Greek prefixes (mono, di, tri,). If the ligand contains a Greek prefix (like ethylenediamine) then use bis (for 2 of them) and tris (for 3 of them).

5. A Roman numeral shows the charge on the central metallic ion.

6. If the complex ion is negative, then the metal name ends with -ate.

7. The preferred order of naming ligands is alphabetical.

22   Review of Selected Topics

Examples:
1.

                      [complex cation   ] [anion]
[Cr(NH$_3$)$_4$Cl$_2$]Cl   dichlorotetraamminechromium III   chloride
                  [2 Cl$^-$]   [4 NH$_3$]   [Cr^{3+}]   [Cl^{1-}]

Compounds with the same molecular formula but with different structures are called isomers. The diagram shows geometrical arrangements of ligands in two octahedral complexes with different physical and chemical properties.

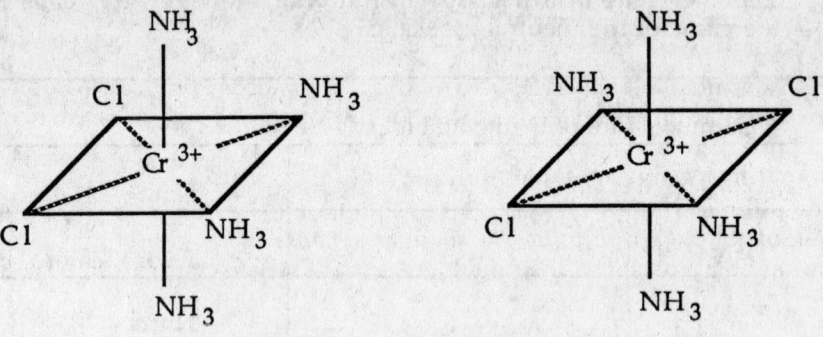

              Cis isomer   Trans isomer

2.

                  [cation] [complex anion]
K$_2$[PtCl$_4$]   potassium tetrachloroplatinate II
              [K^{1+}]   [4 Cl$^-$]   [Pt^{2+}]

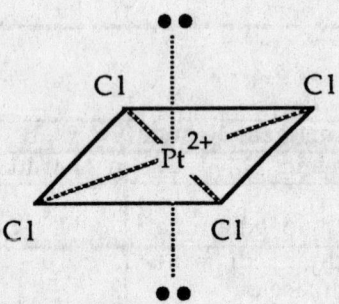

Square Planar

Chapter 3
Stoichiometry

1. The Mole Concept and Molar Mass

A mole is a number, roughly 6 hecto (10^2) giga (10^9) tera (10^{12}) particles. The use of the mole concept in chemistry allows the massing of substances to be substituted for the impossible task of counting atoms, molecules and formulas. A molar mass is the weight, in grams, of Avogadro's number of particles (6.02×10^{23}). Molar mass is computed by the use of atomic weights obtained from the periodic table of the elements.

Examples: The mole concept.

1. A flask contains an unspecified number of krypton (Kr) molecules weighing 21.0 g. What weight of helium molecules contains an equal number of molecules?

    Mole of krypton molecules = $\dfrac{21.0 \text{ g}}{83.8 \text{ g mole}^{-1}}$ = 0.251 mole

    Mole of krypton molecules = mole of helium molecules
    Weight = 0.251 mole × 4.00 g mole^{-1} = **1.00 g**

2. Which 1.0 g sample contains the greatest number of atoms?

    (A) Mercury (Hg)
    mole of atoms = $\dfrac{1.0 \text{ g} \times 1 \text{ atom formula}^{-1}}{201 \text{ g mole}^{-1}}$ = 0.0050 mol

    (B) Carbon dioxide (CO_2)
    mole of atoms = $\dfrac{1.0 \text{ g} \times 3 \text{ atoms molecule}^{-1}}{44.0 \text{ g mole}^{-1}}$ = 0.068 mol

    (C) Zinc sulfate ($ZnSO_4$)
    mole of atoms = $\dfrac{1.0 \text{ g} \times 6 \text{ atoms formula}^{-1}}{162 \text{ g mole}^{-1}}$ = 0.037 mole

    (D) Butane (C_4H_{10})
    mole of atoms = $\dfrac{1.0 \text{ g} \times 14 \text{ atoms molecule}^{-1}}{58.0 \text{ g mole}^{-1}}$ = 0.24 mole

    (E) Hydrogen (H_2) **Answer**
    mole of atoms = $\dfrac{1.0 \text{ g} \times 2 \text{ atoms molecule}^{-1}}{2.0 \text{ g mole}^{-1}}$ = 1.0 mole

The mole is a very convenient unit. A chemical formula gives the relationship between the moles of each element of which it is comprised. Once the moles of compound or the moles of one of the elements in a compound are known, the moles or grams of all of the component parts of the compound can be computed. This is a 'mole link'.

The formula CO_2 could be read to mean that there are:

....twice as many moles of oxygen as mole of carbon. ($n_{oxygen} = 2\, n_{carbon}$)
....half as many moles of carbon as mole of oxygen. ($n_{carbon} = \frac{1}{2}\, n_{oxygen}$)
....three times as many moles of atoms as moles of molecules. ($n_{atoms} = 3\, n_{molecules}$)

Familiarity and comfort using the 'mole link' will go a long way toward solving the common problems that use this concept. It is used to convert from moles of one element to moles of another in a compound, or the moles of the compound from the moles of one of its elements. Just remember to start the process by determining the moles of the substance whose identity and amount is given.

Examples: 'Mole link' in formulas.

1.

Formula	Needed	Known	Relationship
K_2SO_4	oxygen	potassium	n_O = $\frac{4}{2} n_K$ = $2 n_K$
	sulfur	oxygen	n_S = $\frac{1}{2} n_O$ = $\frac{1}{4} n_O$
	potassium	sulfur	n_K = $\frac{2}{1} n_S$ = $2 n_S$
$Na_2C_8H_4O_4$	carbon	sodium	n_C = $\frac{8}{2} n_{Na}$ = $4 n_{Na}$
	sodium	hydrogen	n_{Na} = $\frac{2}{4} n_H$ = $\frac{1}{2} n_H$
	oxygen	carbon	n_O = $\frac{4}{8} n_C$ = $\frac{1}{2} n_C$

2. Element X reacts with oxygen to produce a pure sample of X_4O_{10}. Find the atomic weight and the identity of X if an experiment reacting oxygen with 1.00 g of X resulted in 2.29 g of X_4O_{10}.

$$\text{Mole of oxygen} = \frac{(2.29 - 1.00) \text{ g}}{16.0 \text{ g mole}^{-1}} = \frac{1.29 \text{ g}}{16.0 \text{ g mole}^{-1}} = 0.0806 \text{ mole}$$

$$n_X = \frac{4}{10} \, n_{oxygen} = 0.40 \times 0.0806 \text{ mole} = 0.0323 \text{ mole}$$

$$\text{Molar mass} = \frac{1.000 \text{ g}}{0.0323 \text{ mole}} = 31.0 \text{ g mole}^{-1}$$

The element could be phosphorus, P.

2. Empirical and Molecular Formulas

The ratio of combined atoms in a given compound cannot be changed. To do so would be to describe a different substance. The simplest whole-number ratio of atoms is the 'empirical formula'.

The procedure to compute the empirical formula is:

(a) If the composition (either weights or weight percent) of a compound is known, the moles of each element is calculated.

(b) The mole ratio is determined by dividing the moles of each element of the compound by the smallest number of moles.

(c) An adjustment may be required to adjust the ratio to a whole number ratio (empirical formula).

The 'molecular formula' is a simple integer number 'multiple' (including one) of empirical formulas. The molecular weight is required, and the multiple is the result of dividing the molecular weight by the weight of the empirical formula.

Example

1. Analysis of an organic compound revealed the composition by mass: 22.8% carbon, 1.4% hydrogen, and 75.8% bromine. The molar mass was determined to be 421.7 g mole^{-1}.

   What is the empirical and molecular formula of the compound?

   (a)

Element	Weight	Moles	Ratio	Whole #
carbon	22.79 g	1.90	2.00	4
hydrogen	1.42 g	1.42	1.50	3
bromine	75.79 g	0.948	1.00	2

   (b) The empirical formula is $C_4H_3Br_2$.

   (c) The empirical weight is 210.8.

   (d) The molar mass, 421.7 g mole^{-1}, is twice the empirical weight: (421.7/210.8 = 2.000).

   (e) The molecular formula is $(C_4H_3Br_2)_2$ or $C_8H_6Br_4$.

2. When 1.000 g of an organic compound with the formula $C_xH_yO_z$ is burned in excess oxygen, 1.667 g of carbon dioxide (CO_2) and 0.545 g of water vapor (H_2O) are obtained.

   What is the empirical formula of the compound?

   (a) $n_{carbon\ dioxide}$ = $\dfrac{1.667\ g}{44.0\ g\ mole^{-1}}$ = 0.0379 mole

   (b) $n_{water\ vapor}$ = $\dfrac{0.545\ g}{18.0\ g\ mole^{-1}}$ = 0.0303 mole

   (c) The moles of carbon are equal to the moles of CO_2.
   n_C = 1 x 0.0379 = 0.0379 mole

   (d) The moles of hydrogen are twice the moles of H_2O.
   n_H = 2 x 0.0303 = 0.0606 mole

   (e) The weights of carbon and hydrogen in the compound are computed by multiplying the moles by the molar mass.

   wt_C = 0.0379 mol x 12.0 g mole^{-1} = 0.455 g

   wt_H = 0.0606 mol x 1.0 g mole^{-1} = 0.061 g

(f) The weight of oxygen is found by subtracting the weight of carbon and hydrogen from the total weight of 1.000 g of starting compound.
$wt_O = 1.000 - (0.455 + 0.061) = 0.484$ g
The moles of oxygen is then computed.

Element	Moles	Weight	Moles	Ratio	Whole #
carbon	0.0379	0.455 g	0.0379	1.25	5
hydrogen	0.0606	0.061 g	0.0606	2.00	8
oxygen		0.484 g	0.0303	1.00	4
Total		1.000 g			

Answer: The empirical formula would be $C_5H_8O_4$.

3. **Reaction Stoichiometry**

The coefficients in a balanced chemical reaction equation are directly proportional to the number of moles. The equation

$$H_2(g) + Cl_2(g) \Rightarrow 2HCl(g)$$

can be read to say: "One mole of hydrogen reacts with one mole of chlorine to give two moles of hydrogen chloride".

There is a 'mole link' between the substances in a reaction and this permits the use of a procedure similar to that used in the last section to determine the relationship between the elements in a compound. The 'mole link' allows the weight or the moles of any substance in a chemical equation to be computed when either the weight or the moles of any other chemical in the reaction is given.

This type of problem assumes that:
(a) The reaction proceeds 100% to completion.
(b) There is an 'excess' of the other reactants to produce the amount predicted by consuming the weight of the known substance.

Example

When 9.53 g of carbon disulfide is reacted with an excess of oxygen, carbon dioxide and sulfur dioxide are formed. Find the mass of sulfur dioxide formed in this reaction.

(a) Write and balance the reaction equation.
$$CS_2(g) + 3O_2(g) \Rightarrow CO_2(g) + 2SO_2(g)$$

(b) Compute the moles of the given substance.
$n_{carbon\ disulfide} = \dfrac{9.53\ g}{76.2\ g\ mole^{-1}} = 0.125$ mole

(c) Use the 'mole link' to determine the moles of the product of the reaction.
$n_{sulfur\ dioxide} = {}^2\!/_1\ n_{carbon\ disulfide} = 2 \times 0.125$
$n_{sulfur\ dioxide} = 0.250$ mole

(d) Determine the weight of a substance.
$wt_{sulfur\ dioxide} = 0.250$ mole $\times\ 96.1\ g\ mole^{-1} = 24.0$ g

4. Limiting Reactants

Reaction problems are given where the quantities of <u>two</u> (rarely more than two) reactants are stated. The problem allows the assumption of 100% reaction for one of the substances. It is then necessary to determine which of the two would be used up...the 'limiting' reactant. The 'limiting' reactant then would be used with the 'mole link' to compute the amount of product produced.

Example

The reaction of chromium with sulfur produces chromium(III) sulfide. How many grams of chromium(III) sulfide can be isolated when 26.0 g of chromium is reacted with 77.0 g of sulfur?

(a) Write and balance the reaction equation.

$$16Cr(s) + 3S_8(s) \Rightarrow 8Cr_2S_3(s)$$

(b) Compute the moles of the given substances.

$$n_{chromium} = \frac{26.0 \text{ g}}{52.0 \text{ g mole}^{-1}} = 0.500 \text{ mole}$$

$$n_{sulfur} = \frac{77.0 \text{ g}}{257 \text{ g mole}^{-1}} = 0.300 \text{ mole}$$

(c) Use the 'mole link' to determine the 'limiting reactant'.

$$n_{chromium} = {}^{16}\!/\!_3 \; n_{sulfur} = {}^{16}\!/\!_3 \times 0.300 = 0.533 \text{ mole}$$
This exceeds the amount given.
Chromium <u>is</u> the 'limiting reactant'.

$$n_{sulfur} = {}^{3}\!/\!_{16} \; n_{chromium} = {}^{3}\!/\!_{16} \times 0.500 = 0.0938 \text{ mole}$$
This does not exceed or equal the amount given.
Sulfur is not the 'limiting reactant'.

(d) Determine the weight of the product.

$$n_{chromium\,sulfide} = {}^{8}\!/\!_{16} \; n_{chromium} = \tfrac{1}{2} \times 0.500$$
$$n_{chromium\,sulfide} = 0.250 \text{ mole}$$

$$wt_{chromium\,sulfide} = 0.250 \text{ mole} \times 200. \text{ g mole}^{-1}$$
$$wt_{chromium\,sulfide} = 50.0 \text{ g}$$

5. Molarity, Molality and Mole Fraction

A solution is a homogeneous mixture of a solute and a solvent.

(Solute + Solvent = Solution).

Many reactions occur in solution, particularly in water solution. The most common concentration used in chemistry to indicate the quantity of solute dissolved in a given amount of solvent is molarity.

$$\text{Molarity (M)} = \frac{\text{moles solute (n)}}{\text{Liter solution (V)}}$$

The freezing point depression and the boiling point elevation measurements of a solution is frequently used to determine the molecular weight or the percent ionization of a solute. The temperature change depends upon the number of molecules or ions dissolved in a specified amount of solvent, and the preferred concentration unit is molality.

$$\text{Molality (m)} = \frac{\text{moles solute (n)}}{\text{kilogram solvent (kg)}}$$

Solutions of a nonvolatile solute in a liquid solvent will have a lower vapor pressure than the pure solvent. Raoult's Law allows the calculation of the vapor pressure of a solution by multiplying concentration by the vapor pressure of the solvent. The concentration used is mole fraction.

$$\text{Mole fraction (x)} = \frac{\text{moles solute}}{\text{moles solution}}$$

Note that:

(a) The numerator in the three concentration expressions is the same. Converting from one to another requires adjusting the denominator.

(b) The density of the solution is needed for conversions between molarity and molality. The molecular weight of the solvent is used to convert between mole fraction and molality.

(c) It is inconvenient to change between molarity and mole fraction directly, so molality should be calculated as an intermediate step.

$$\text{Molarity (M)} \quad \overset{\text{density}}{\Leftrightarrow} \quad \text{Molality (m)} \quad \overset{MW_{solvent}}{\Leftrightarrow} \quad \text{Mole fraction (x)}$$

(d) For the same solution, molality always has the greatest numerical value, and mole fraction the lowest.

$$\text{Mole fraction (x)} \quad << \quad \text{Molarity (M)} \quad < \quad \text{Molality (m)}$$

Stoichiometry

Example

1. The density of a solution of 20.0 % (by weight) KI (FW = 166.03) solution in water is 1.166 g/mL (1.166 kg/L).

    Basis: 1000. g$_{solution}$ = 200. g$_{solute}$ + 800. g$_{solvent}$
             1.000 kg$_{solution}$ = 0.200 kg$_{solute}$ + 0.800 kg$_{solvent}$

    (a) Calculate the molarity.   (Ans. **1.40 M**)

    n_{solute} = $\dfrac{200. \text{ g}}{166 \text{ g mole}^{-1}}$ = 1.20 mole

    $L_{solution}$ = $\dfrac{1.000 \text{ kg}_{solution}}{1.166 \text{ kg L}^{-1}}$ = 0.858 L

    Molarity = $\dfrac{1.20 \text{ mole}}{0.858 \text{ L}}$ = **1.40 M**

    (b) Calculate the molality.   (Ans. **1.50 m**)

    n_{solute} = 1.20 mole   (from part (a))
    $kg_{solvent}$ = 0.800 kg
    Molality = $\dfrac{1.20 \text{ mole}}{0.800 \text{ kg}}$ = 1.50 m

    (c) Calculate the mole fraction.   (Ans. **0.0263**)

    n_{solute} = 1.20 mole   (from part (a))
    $n_{solvent}$ = $\dfrac{800. \text{ g}_{solvent}}{18.0 \text{ g mole}^{-1}}$ = 44.4 mole

    $n_{solution}$ = 44.4 mole$_{solvent}$ + 1.20 mol$_{solute}$ = 45.6 mole

    mole frac. = $\dfrac{1.20 \text{ mole}}{45.6 \text{ mole}}$ = 0.0263

2. A 1.00 molal solution of NaOH(aq) has a density of 1.032 g mL^{-1}.

    (a) Calculate the molarity.   (Ans. **0.992 M**)

    n_{solute} = 1.00 mole   (from 1.00 molal)
    $L_{solution}$ = $\dfrac{(1.000 \text{ kg}_{solvent} + 0.040 \text{ kg}_{solute})}{1.032 \text{ kg L}^{-1}}$ = 1.01 L

    Molarity = $\dfrac{1.00 \text{ mole}}{1.01 \text{ L}}$ = **0.992 M**

    (b) Calculate the mole fraction.   (Ans. **0.0177 M**)

    n_{solute} = 1.00 mole   (from 1.00 molal)
    $n_{solvent}$ = $\dfrac{1000 \text{ g}_{solvent}}{18.0 \text{ g mole}^{-1}}$ = 55.6 mole

    $n_{solution}$ = 55.6 mole$_{solvent}$ + 1.00 mol$_{solute}$ = 56.6 mole

    mole frac. = $\dfrac{1.00 \text{ mol}_{solute}}{55.6 \text{ mol}_{solution}}$ = 0.0177

6. Solution Stoichiometry

The labels on reagent bottles gives the molarity ($\underline{M}$) of each solution used in a reaction. The moles of solute are calculated using the molarity and volume of each solution in the reaction (moles = Molarity x Volume). The coefficients of the balanced reaction equation provide a 'mole link' which can be used to compute the moles of a substance that react with or produced from a known amount of a reactant.

Example

When 20.0 mL of 0.15 $\underline{M}$ magnesium sulfate is reacted with a 0.10 $\underline{M}$ sodium hydroxide solution, magnesium hydroxide is precipitated. Find the minimum amount of NaOH(aq) required to completely react with the $MgSO_4$(aq); and the maximum yield of dried precipitate that would be expected.

(a) Write and balance the reaction equation.
Mg^{2+}(aq) + $2OH^-$(aq) $\Rightarrow$ $Mg(OH)_2$(s)

(b) Compute the moles of the given substance.
$n_{Mg^{2+}}$ = 0.15 mole L^{-1} x 0.020 L = 0.0030 mole

(c) Use the 'mole link' to determine the moles of the reactant required.
n_{OH^-} = $^2/_1$ $n_{Mg^{2+}}$ = 2 x 0.0030
n_{OH^-} = 0.0060 mole
$n_{Mg(OH)_2}$ = $n_{Mg^{2+}}$ = 0.0030 mole

(d) Determine the volume of reactant needed and the weight of product which will be produced.
mL_{NaOH} = $\dfrac{0.0060 \text{ mole}}{0.10 \text{ mole } L^{-1}}$ = 0.060 L = 60. mL

$wt_{Mg(OH)_2}$ = 0.0030 mole x 58.3 g $mole^{-1}$ = **1.75 g**

7. Colligative Properties

Freezing point depression (ΔT_f), boiling point elevation (ΔT_b) and vapor pressure lowering ($\Delta P°$) depend on the number of solute particles (ions or molecules) present in the solvent. The lower vapor pressure is caused by the attractions between solute and solvent molecules. The freezing point and boiling point of the solution is dependent upon vapor pressure. The result is a depression of the freezing point and an elevation of the boiling point of the solution when compared to those of the solvent. The units of choice for concentration that are used in colligative property equations are either molality (m) or mole fraction (x).

K_f and K_b are the symbols for the molal freezing and boiling point constants, and these values are always supplied.

The key formulas are:

fp and bp change	$\Delta T_f = K_f m$	$\Delta T_b = K_b m$
Raoult's Law	$P_{solvent} = x_{solvent} P°$	$\Delta P_{solvent} = x_{solute} P°$

Example:

1. In a melting–temperature determination to approximate the molecular weight of a compound, 1.005 g of the compound is dissolved in 12.01 g of naphthalene. The freezing point was lowered by 4.4°C. The freezing point constant (K_f) for naphthalene is 6.6° molal^{-1}. What is the approximate molecular weight of the compound?

   (a) Determine the weight of solute per kg solvent.
   $$g_{solute}\,kg^{-1} = \frac{1.01\ g_{solute}}{12.07\ g_{solvent}} \times 1000\ g\ kg^{-1}$$
   $$g_{solute}\,kg^{-1} = 83.7\ g\ kg^{-1}$$

   (b) Determine the molality of the solution.
   $$\text{molality} = \frac{\Delta T_f}{K_f} = \frac{4.4\ °C}{6.6\ °C\ m^{-1}} = 0.67\ \text{mole}\ kg^{-1}$$

   (c) Determine the molecular weight of the solute.
   $$MW = \frac{83.7\ g\ kg^{-1}}{0.67\ \text{mole}\ kg^{-1}} = 130\ g\ \text{mole}^{-1}$$

2. The vapor pressure of toluene is 28.829 Torr. A solution of a nonvolatile, nonionizing solute in toluene is made. When 4.256 g compound is dissolved in 100. g of toluene (C_7H_8, FW = 92.07) the vapor pressure is 28.209 Torr. What is the approximate molecular weight of the solute?

   (a) Find the mole fraction of solvent from Raoult's Law.
   $$x_{solute} = \frac{\Delta P_{solvent}}{P°_{solvent}} = \frac{(28.829 - 28.209)}{28.829} = 0.0215$$
   $$x_{solvent} = 1.0000 - 0.0215 = 0.9785$$

   (b) Determine the moles of solvent, solute and solution.
   $$n_{solvent} = \frac{100.0\ g}{92.07\ g\ \text{mole}^{-1}} = 1.086\ \text{mole}$$
   $$n_{solute} = \frac{4.256\ g}{MW\ g\ \text{mole}^{-1}} = 4.256/MW$$
   $$n_{solution} = 1.086 + (4.256/MW)$$

   (c) Calculate molecular weight using the mole fraction definition.
   $$x_{solvent} = 0.9785 = \frac{1.086}{1.086 + (4.256/FW)}$$
   $$MW = 178\ g\ \text{mole}^{-1}$$

Chapter 4
States of Matter

1. The Solid State

In the process of melting or subliming, the atoms, ions or molecules which make up a solid become separated. The more strongly the units of a lattice attract each other, the greater is the temperature required to separate them.

High melting crystalline solids are held together by metallic, ionic or network covalent bonds. The few substances that form network covalent bonds include carbon in the form of diamond or graphite and silicon dioxide (SiO_2) in the from of quartz, mica or asbestos.

Molecules are held in crystal lattices by attractive forces. The melting temperatures of these substances are significantly lower than those of metals and salts.

<u>Hydrogen bonding</u> is the strongest intermolecular force of attraction. A hydrogen atom which is covalently bonded to a small, highly electronegative atom is also attracted to the unshared electron–pair of the electronegative atom in a nearby molecule. This type of attraction always involves molecules containing hydrogen atoms (H) bonded to nitrogen (N), oxygen (O) or fluorine (F). The hydrogen nucleus is both electron–deficient and unshielded by inner shell electrons, and makes up for it by sharing the lone–pair electrons on the nearby molecule containing nitrogen, oxygen or fluorine.

<u>Van der Waals' forces</u> are the weakest attractive forces between molecules. There are two major van der Waals' forces.

1. <u>Dipole attractions</u> occur when polar molecules are attracted to one another.

2. <u>Dispersion forces</u> are the weakest of all molecular attractions, and are caused by instantaneous, temporary dipoles caused by the motion of electrons. The strength of dispersion forces increases as the number of electrons in a molecule increases. This force is now called a 'London Force'.

<u>Example</u>
State which has an ionic, a molecular covalent, or a network covalent structure.

	Formula	Data and properties	Structure
1.	HBr	mp -88.5 °C bp - 67 °C	Molecular
2.	BN	sublimes at 3000°C, all single bonds	Network
3.	CS_2	mp -110.8 °C bp 46.3 °C double covalent bonds between C=S	Molecular
4.	$MgCl_2$	mp 708 °C bp 1412 °C conducts electricity when molten	Ionic
5.	$AsCl_3$	mp - 8.5 °C bp 130.2 °C	Molecular

2. Changes of State

A graph of pressure plotted against temperature displays the regions of all the phases of a pure substance at the same time. The phase change diagram for benzene, C_6H_6, shows several characteristics which can be illustrated by this type of presentation.

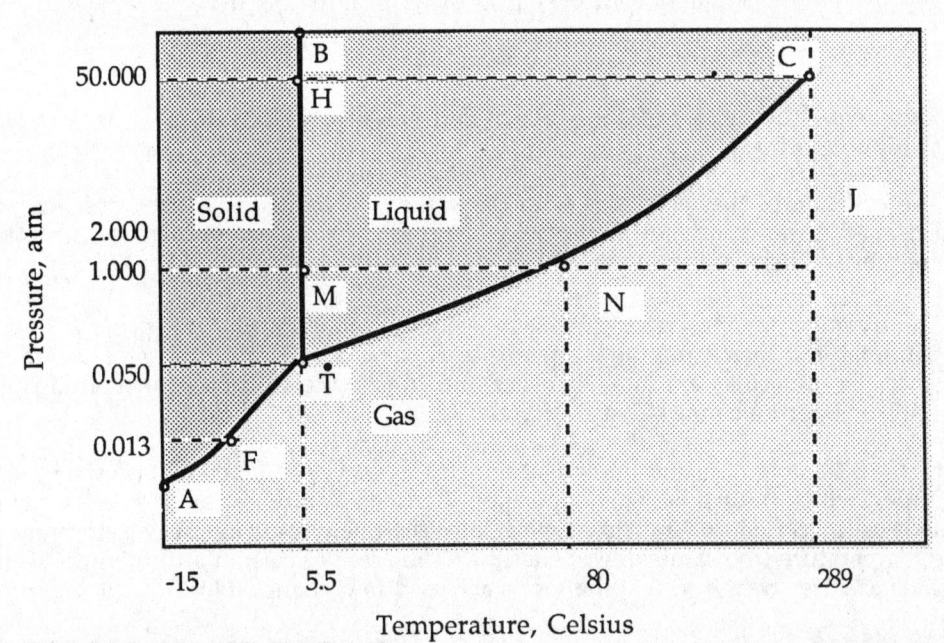

The equilibrium <u>vapor pressure</u> is the pressure of the vapor in equilibrium with a liquid or solid at a given temperature. It is dependent upon temperature only, and is represented in the diagram by segments 'TC' and 'TA'.

<u>Examples</u> of the use of the phase change diagram.

1. The <u>sublimation pressure</u> curve is segment TA.

    At a pressure of 10 mm Hg (0.013 atm), benzene will sublime (go from the solid state directly to the gaseous state) at the temperature corresponding to point 'F' (about 11 °C).

2. The <u>vapor pressure</u> curve is TC.

    The normal boiling temperature of benzene is 80 °C. At a pressure of 1 atmosphere, benzene will boil at the temperature corresponding to point 'N' (80 °C).

3. The <u>melting temperature</u> curve is TB.

    The normal melting temperature of benzene is 5.5 °C. At a pressure of 1 atmosphere, benzene will melt at the temperature corresponding to point 'M' (5.5 °C).

4. The <u>triple point</u> is 'T'.

    At a pressure of 36 mm Hg (0.05 atm) and a temperature of 5.5 °C all three states of matter; solid, liquid and gaseous benzene, exist simultaneously.

5. The <u>critical</u> <u>point</u> is 'C'.

This point is at the <u>critical</u> <u>temperature</u> (289°C), above which it is impossible to liquefy a gaseous C_6H_6, no matter how much the pressure is increased.
It also marks the <u>critical</u> <u>pressure</u> (50 atm.) which is the minimum amount required to liquefy benzene gas at the critical temperature.

3. Kinetic Molecular Theory of Gases; Graham's Law

The kinetic theory of gases postulates that gaseous molecules move in a straight-line motion at high velocities.

(1) The collision of the molecules with the walls of a container causes gaseous pressure. Decreasing the volume of the vessel will cause the pressure to increase, since there are are more collisions per square unit of wall area (Boyle's Law).

(2) The average kinetic energy (mass x velocity squared, mv^2) is proportional to the temperature. Increasing the temperature of the gas will cause the pressure to increase, since the molecules are moving faster and hitting the wall at an increased number of collisions per unit time (Charles' Law).

Gases diffuse or mix, and the rate at which this occurs is dependent on the speed of the gas molecules, which in turn is dependent upon the temperature. The escape of a gas through a pin hole is called **effusion**. Graham's Law relates the rate of effusion of two different gases at the same temperature (the same kinetic energy). The rate of effusion and the rate of diffusion of two gases is close enough so that the terms are used interchangeable.

$$\frac{\text{Rate of effusion of Gas A}}{\text{Rate of effusion of Gas B}} = \sqrt{\frac{\text{Molecular Weight of Gas B}}{\text{Molecular Weight of Gas A}}}$$

Example:

1. What is the relative rate of diffusion of CO (FW = 28.0) and CO_2 (FW = 44.0)?
   (A) CO is 1.25 times faster than CO_2
   (B) CO_2 is 1.25 times faster than CO
   (C) CO is 1.57 times faster than CO_2
   (D) CO_2 is 1.57 times faster than CO
   (E) They diffuse at the same rate

The answer is (A).

$$\frac{\text{Rate of effusion of CO}}{\text{Rate of effusion of CO}_2} = \sqrt{\frac{\text{Molecular Weight of CO}}{\text{Molecular Weight of CO}_2}} = \sqrt{\frac{44.0}{28.0}}$$

$$\frac{\text{Rate of effusion of CO}}{\text{Rate of effusion of CO}_2} = 1.25$$

2. Which of the gases effuses more quickly than oxygen, O_2?
   (A) F_2
   (B) N_2O
   (C) H_2S
   (D) NO
   (E) Cl_2

   The answer is (D). NO (FW = 30) is the only choice which has a lower molecular weight than oxygen.

4. **The Ideal Gas Law**

   Gases will take the volume of the vessel in which they are contained, no matter what the pressure or temperature might be. The amount (n, in moles), the volume (V, in Liters), the pressure (P, in atmospheres) and the temperature (T, in Kelvins) are related by the Ideal Gas Law Equation (PV = nRT). R is the universal gas constant. Given any three of the variables in the equation allows the fourth to be computed.

   Real gases, as contrasted to ideal gases, are subject to at least van der Waals' forces of attraction. In calculations requiring high accuracy, a term n^2a/V^2 is added to the pressure, $(P + n^2a/V^2)$, to correct for intermolecular attractions. In addition, the term nb is subtracted from the volume, (V - nb), to correct for that portion of the volume that is not compressible because real gases have an intrinsic volume.

   Examples:

   Rearrangements of the ideal gas law equation allow its use in the calculation of gaseous density and molar mass from experimental data.

1. What is the density of nitrogen gas (g/L) in a sample that exerts a pressure of 150. mm Hg at 0°C.?

$$PV = \left[n\right]RT = \left[\frac{g}{MW}\right]RT$$

$$D = \frac{g}{V} = \frac{P * MW}{RT}$$

$$D = \frac{g}{V} = \frac{(150/760 \text{ atm}) \times (28.0 \text{ g mol}^{-1})}{(0.0821 \text{ Liter–atm mol}^{-1} K^{-1}) \times (273 K)} = 0.247 \text{ g/L}$$

2. 1.135 g of a gaseous compound containing only nitrogen and oxygen occupies 642 mL at a pressure of 720 mm Hg and a temperature of 27.0°C. What is the molecular weight of this gas?

$$PV = \left[\frac{g}{MW}\right]RT$$

$$MW = \frac{gRT}{PV}$$

$$MW = \frac{1.135 \text{ g} \times 0.0821 \text{ Liter–atm mol}^{-1} K^{-1} \times 300 K}{(720/760 \text{ atm}) \times (642/1000 \text{ L})} = 46.0 \text{ g mol}^{-1}$$

5. Dalton's Law of Partial Pressures

Each gas in a mixture exerts a partial pressure as if it were present alone. This pressure is proportional to the number of moles of the gas present.

Example:

A mixture of methane, CH_4, and ethyne, C_2H_2, occupied a certain volume at a total pressure of 100. mm Hg. The sample was burned, and the CO_2 alone was collected. The pressure of the CO_2 was found to be 150. mm Hg at the same temperature as the original mixture. What fraction of the gas was CH_4?

$$CH_4(g) + 2O_2(g) \Rightarrow CO_2(g) + H_2O(l)$$
$$C_2H_2(g) + \tfrac{5}{2} O_2(g) \Rightarrow 2CO_2(g) + H_2O(l)$$

Of the CO_2 that resulted from the combustion, ⅓ came from the methane, and ⅔ came from the ethene.

Final mixture:
⅓ × 150. = 50. mm Hg is the partial pressure of the CH_4
⅔ × 150. = 100. mm Hg is the partial pressure of the C_2H_2

Initial mixture:
The partial pressure of the CH_4, = ⅓ × 100. = 33 mm Hg
The partial pressure of the C_2H_2, = ⅔ × 100. = 67 mm Hg
The fraction of the gas that was CH_4 is ⅓ or 0.33.

6. Properties of Solutions; Raoult's Law

The vapor pressure lowering of a solution is proportional to the mole fraction of nonvolatile solute dissolved in the solvent.

$\Delta P = x_{solute} P°_{solv}$
   x_{solute} is the mole fraction of solute
   $P°_{solv}$ is the vapor pressure of the pure solvent

It is convenient to have a Raoult's Law expression in terms involving only the solvent.

$P_{solv} = x_{solv} P°_{solv}$
   x_{solv} is the mole fraction of solvent
   P_{solv} is the vapor pressure of the mixture

If the solution is made up of two volatile solvents, the partial pressure of each of the solvents in the vapor above the solution is calculated from:

$P_{solv1} = x_{solv1} P°_{solv1}$
$P_{solv2} = x_{solv2} P°_{solv2}$

and the total pressure of the vapor is:

$P_{Total} = P_{solv1} + P_{solv2}$
$P_{Total} = x_{solv1} P°_{solv1} + x_{solv2} P°_{solv2}$

Examples:

1. What is the vapor pressure at 34.9°C of 184 g of ethanol, C_2H_5OH, in which 92.1 g of glycerol, $C_3H_5(OH)_3$, is dissolved. The vapor pressure of pure ethanol at 34.9°C is 100.0 mm Hg. Glycerol is nonvolatile at 34.9°C.

$$x_{ethanol} = \frac{(184.0 \text{ g} \div 46.0 \text{ g mol}^{-1})}{(92.1\text{g} \div 92.1 \text{ g mol}^{-1}) + (184.0 \div 46.0)} = 0.800$$

$P_{solv} = x_{solv} P°_{solv} = 0.800 \times 100.0 \text{ mm Hg} = 80.0 \text{ mm Hg}$

2. The vapor pressure of benzene, C_6H_6, and toluene, C_7H_8, are 95.1 mm Hg and 28.4 mm Hg, respectively.
What is the composition of the vapor above an equimolar solution of these solvents ($x_{benzene} = x_{toluene} = 0.500$) ?

$P_{benzene}$	$P_{toluene}$
$= x_{benzene} P°_{benzene}$	$= x_{toluene} P°_{toluene}$
$= 0.500 \times 95.1$	$= 0.500 \times 28.4$
$= 47.6$ mm Hg	$= 14.2$ mm Hg
$P_{Total} = 47.6 + 14.2 = 61.8$ mm Hg	

Dalton's law of partial pressures predicts that the ratio of each partial pressure to the total pressure is proportional to the moles of that component.

$x_{benzene}$	$x_{toluene}$
$= P_{benzene} \div P_{Total}$	$= P_{toluene} \div P_{Total}$
$= 47.6 \div 61.8 = 0.770$	$= 14.2 \div 61.8 = 0.230$

Answer: The vapor is 77.0% benzene and 23.0% toluene by volume or by moles.

Chapter 5
Reaction Kinetics

1. Rate Equations

The rate of a chemical reaction depends on the frequency and force of collisions between molecules. Changes in the concentration and the temperature of the system are observed experimentally to establish a rate equation for the reaction.

For the hypothetical reaction:

$$2A + 4B \Rightarrow 5C + 3D$$

$$\text{Rate} = -\frac{1}{2}\frac{\Delta[A]}{\Delta t} = -\frac{1}{4}\frac{\Delta[B]}{\Delta t} = +\frac{1}{5}\frac{\Delta[C]}{\Delta t} = +\frac{1}{3}\frac{\Delta[D]}{\Delta t}$$

(Note: Δ represents 'change in'):

Two (2) A's are reacting, and A is disappearing as time goes on. Because A is decreasing, a minus sign is used in the rate equation. Half the rate of disappearance of A with time will give the specific rate of reaction. 5 C's are being made as a product, so ⅕ the rate of appearance (plus sign) of C will give the same rate for the reaction.

For the hypothetical reaction:

$$3A + B \Rightarrow 2C + D$$

$$\text{Rate} = -\frac{1}{3}\frac{\Delta[A]}{\Delta t} = +\frac{1}{2}\frac{\Delta[C]}{\Delta t} = k[A]^m[B]^n$$

The order of a reaction is determined experimentally and is dependent on reactant concentration. It is shown as an exponent of the molarity, and is generally a simple whole number. Sometimes it might be a square root.

The order may or may not be the same as the stoichiometric coefficients of the balanced equation. Generally they will not be the same. In the sample reaction:

    m is the "order" with respect to A
    n is the "order" with respect to B
    m + n is the overall "order"

The specific rate constant, k, is dependent on reaction temperature, and must be determined experimentally.

Examples

1. Given the reaction: $N_2O_3(g) \Rightarrow NO(g) + NO_2(g)$

Data	Pressure $_{(N_2O_3)}$	Rate
Run 1	0.0015 atm	0.00917 atm sec^{-1}
Run 2	0.0020	0.0122
Run 3	0.0025	0.0153
Run 4	0.0030	0.0183

   1. The rate equation is determined by setting up a ratio of the concentration and seeing what power it must be to fit the rate data. It makes things simpler to look for a doubling of concentration or pressure.

      $$\frac{[0.0030]^m}{[0.0015]} = \frac{[0.0183]}{[0.00917]}$$

      Since doubling the pressure doubled the rate, m=1, and the reaction is 'first order'.

   2. After the order of the reactants has been determined, it is possible to determine the value of the specific rate constant, k, including its units.

      Rate = k $[P_{N_2O_3}]^1$

      0.00917 atm sec^{-1} = k × 0.0015 atm
      k = 6.11 sec^{-1}

2. Given the hypothetical reaction: A + 2B + C $\Rightarrow$ Products

Data	[A]	[B]	[C]	Time
Run 1	0.20 M	0.20 M	0.20 M	160 sec
Run 2	0.20	0.40	0.10	160
Run 3	0.40	0.40	0.10	40
Run 4	0.40	0.40	0.20	20

   1. To write the rate equation, first note that time is given. The rate is the reciprocal of time.

      Get the order of A using Run 2 and Run 3, since [B] and [C] are not changing.

      $$\frac{[0.40]^m}{[0.20]} = \frac{[1/40]}{[1/160]} = \frac{[160]}{[40]} = 4.0 \quad \text{The order of A is 2.}$$

      Get the order of C using Run 3 and Run 4, since [A] and [B] are not changing.

      $$\frac{[0.20]^n}{[0.10]} = \frac{[1/20]}{[1/40]} = \frac{[40]}{[20]} = 2.0 \quad \text{The order of C is 1.}$$

Get the order of B using Run 1 and Run 2, since [A] is not changing and we know the order of C.

$$\frac{[0.40]^r [10]^1}{[0.20] \ [20]} = \frac{[1/160]}{[1/160]} = 1.0 \qquad \text{The order of B is 1.}$$

The overall order is 4, and the rate equation is:

$$\text{Rate} = k[A]^2[B]^1[C]^1$$

2. The value of the rate constant, k, can be determined by using Run 1 and the rate of disappearance of A.

The units of k will be the reciprocals of time and molarity (time^{-1}; molarity^{-1}). The exponent of the molarity is one less than the overall order of 4.

$$\frac{0.20 \text{ M}}{160 \text{ sec}} = k[0.20M]^2[0.20M]^1[0.20M]^1$$

$$k = 0.78 \text{ liter}^3 \text{ mole}^{-3} \text{sec}^{-1}$$

2. First Order Reactions

The half–life relationship used with nuclear reactions is an example of a first–order reaction. When the rate of a chemical reaction is found to be first–order, the same half–life relationship exists, and the same equations can be used.

$$\ln \frac{[N]}{[N_o]} = -kt \qquad k = \frac{0.693}{t_{1/2}}$$

The half–life is the amount of time required for half of a given amount of reactant to be consumed, and is independent of the initial concentration of the reactant.

Example:
The decomposition of hydrogen peroxide, H_2O_2, at 70°C. is first–order and has a rate constant of 0.0347 min^{-1}

$$2H_2O_2(aq) \Rightarrow 2H_2O(l) + O_2(g)$$

How long will it take for 70.% of a sample of hydrogen peroxide to decompose (30.% remains), and what is the half–life of H_2O_2 at 70°C?

$$\ln \frac{[N]}{[N_o]} = -kt$$

$$\ln 0.30 = 0.0347 \text{ min}^{-1} \times t$$

Answer: t = 35 minutes

$$t_{1/2} = \frac{0.693}{k}$$

$$t_{1/2} = \frac{0.693}{0.0347 \text{ min}^{-1}}$$

Answer: $t_{1/2}$ = 20 minutes

3. Activation Energy and Catalysis

The collision theory of reaction rates describes reactions in terms of collisions between reacting molecules. To be effective, a collision must be both aligned properly and energetic enough to form an intermediate species called the 'activated complex'. The number of effective collisions is a small fraction of the total number of collisions between reacting molecules. Most of the molecules bounce off one another and remain unchanged.

The 'activation energy', E_a, is the minimum energy required to cause a reaction between molecules colliding with the proper geometry.

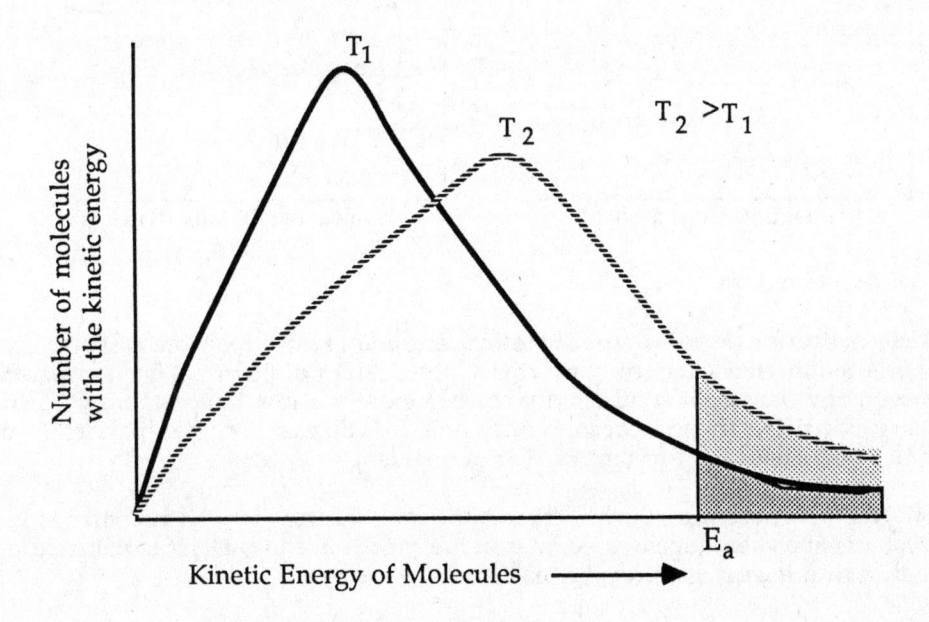

The kinetic energy diagram illustrates the one reason why the often cited small increase in temperature will greatly increase the rate of reaction. The number of molecules that have at least the activation energy, E_a, has more than doubled.

A potential energy diagram shows how the energy of the reactants, activated complex and products are related.

The diagram on the left is for an exothermic reaction. The figure to the right is for an endothermic reaction.

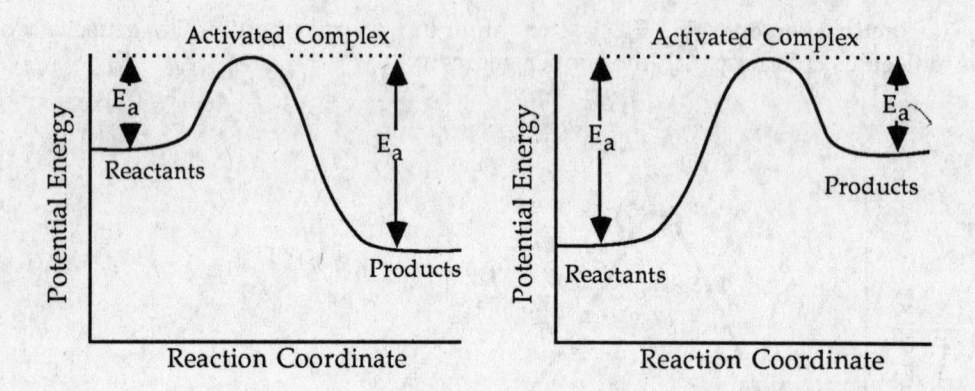

The two activation energies are for the forward and reverse reaction, respectively. The kinetic energy of the reactants may be converted to the amount of potential energy required by the activation energy. High potential energy corresponds to a low bond stability. The high potential energy of the activated complex indicates it will easily come apart to form the products. This is the fastest part of the reaction mechanism.

A catalyst is a substance which increases the rate of a reaction. It is consumed in one step of the reaction and then regenerated later in the process. The catalyst is not used up, but provides a new, lower energy path for the reaction.

The diagrams show the uncatalyzed and catalyzed path for the same exothermic reaction.

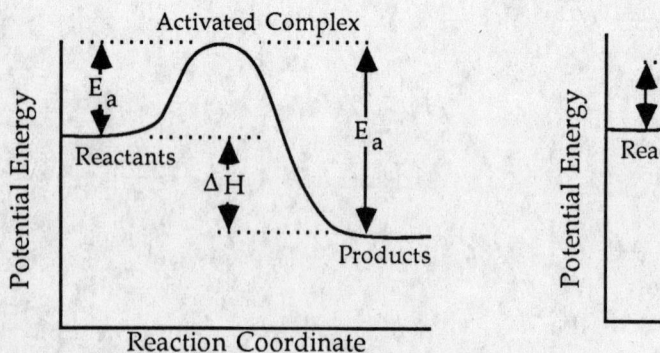

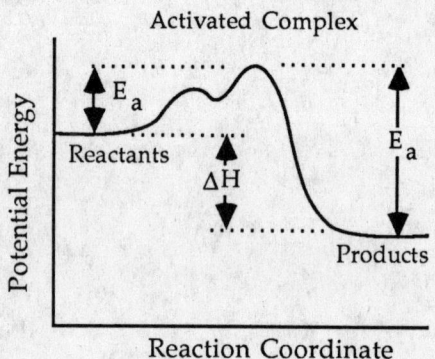

The catalyzed reaction has a lower activation energy, and there are significantly more molecules with the kinetic energy required to form the activated complex.

Two additional points are illustrated by these figures.

1. The heats of reaction (ΔH) for an uncatalyzed and catalyzed reactions are the same. The energy of the reactants and products is the same for both reactions.

2. The step of a catalyzed reaction with the highest activation energy is the rate determining (slowest) step. There are fewer molecules with the required energy.

4. **Heat Effects and the Arrhenius Equation**

The quantitative relationship between the rate constant, k, and temperature, T. The basic relationship was first shown to be valid in 1889 by Svante Arrhenius.

$$k = Ae^{-E_a/RT}$$ where: A is a constant
E_a is the activation energy.

The plot of ln k vs $1/T$ is a straight line, and the slope of this line is equal to $-E_a/R$. You may have done a laboratory experiment that allows the estimation of the activation energy of a reaction. Some texts develop this equation using logs, but modern calculators have both 'log' and natural log (lnx) functions. The use of the Arrhenius equation in the base 'e' form precludes the need to use the conversion between the two log forms (lnx = 2.303 log x).

An integrated version of the Arrhenius equation is used to determine:

1. The value of the rate constant at different temperatures when the activation energy is known.

2. The value of the activation energy when the rate constants at two temperatures are known.

Remember the form of this equation. It is also used to calculate the change with temperature of the equilibrium constant, K, by using the value for the heat of reaction, ΔH, instead of E_a.

$$\ln \frac{k_1}{k_2} = -\frac{E_a}{R}\left[\frac{1}{T_2} - \frac{1}{T_1}\right]$$

The value of the gas constant, R, used to calculate the activation energy is either 1.987 cal mol^{-1} K^{-1} or 8.31 Joule mol^{-1} K^{-1}. Both rate constants must have the same dimensions.

Example:
Given a reaction where:
  k_1 is 5.00×10^{-2} mol^{-1} sec^{-1} at 287°C.
  k_2 is 6.80×10^0 mol^{-1} sec^{-1} at 333°C.

To calculate the activation energy, first write the Arrhenius equation. Then:

$$\ln \frac{6.80 \times 10^0}{5.00 \times 10^{-2}} = \frac{-E_a}{1.987 \text{ cal mol}^{-1} \text{ K}^{-1}} \left[\frac{1}{606} - \frac{1}{560} \right]$$

$$4.91 = \frac{-E_a}{1.987 \text{ cal mol}^{-1} \text{ K}^{-1}} \left[1.65 \times 10^{-3} - 1.79 \times 10^{-3} \right]$$

$E_a = +72{,}000$ cal $= 72{,}000$ cal $\times 4.184$ J cal^{-1}
$E_a = +301{,}000$ J

2. What would be the value of the rate constant at 313°C?

$$\ln \frac{k_2}{5.00 \times 10^{-2}} = \frac{-72{,}000 \text{ cal}}{1.987 \text{ cal mol}^{-1} \text{ K}^{-1}} \left[\frac{1}{586} - \frac{1}{560} \right]$$

$$\ln \frac{k_2}{5.00 \times 10^{-2}} = 2.87$$

A calculator should be used to find a value which is the "inverse ln x" of 2.87 (or "e^x" of 2.87).

$$\frac{k_2}{5.00 \times 10^{-2}} = 17.7$$

$k_2 = 17.7 \times 5.00 \times 10^{-2}$
$k_2 = 8.83 \times 10^{-1}$ mol^{-1} sec^{-1} at 313°C.

5. Reaction Mechanism

A series of steps (usually 2–4) that add up to the overall reaction is called the reaction mechanism.

Principles:

1. OVERALL REACTION: $2 NO_2 + F_2 \Rightarrow 2 NO_2F$
   MECHANISM:
   1. $NO_2 + F_2 \Rightarrow NO_2F + F$ (slow)
   2. $F + NO_2 \Rightarrow NO_2F$ (fast)

   PRINCIPLES INVOLVED:

   a. The *slow step* is the rate–determining step of the overall reaction, and determines the rate law for the overall reaction.
   b. An *intermediate* is formed from the reactants and is involved in the process, but does not appear in the overall reaction. Elemental fluorine, F, is an intermediate in this mechanism.

   RATE EQUATION: Rate = $k_1[NO_2][F_2]$

2. OVERALL REACTION: $2 O_3 \Leftrightarrow 3 O_2$
   RATE EQUATION:

   $$\text{Rate} = \frac{k[O_3]^2}{[O_2]} \quad \text{where } k = K_1 * k_2$$

   MECHANISM:
   1. $O_3 \Leftrightarrow O_2 + O$ (fast equilibrium)
   2. $O_3 + O \Rightarrow 2 O_2$ (slow)

   PRINCIPLE INVOLVED:

   Use the equilibrium step to *remove intermediates* from the rate equation. It is the possible involvement of intermediates that makes it possible to propose these mechanisms, but an intermediate should not appear in the rate law equation.

   A.   Rate = $k_2 [O_3][O]$ (from the slow step)

   B. $K_1 = \dfrac{[O_2][O]}{[O_3]}$ (from the equilibrium step)

         $[O] = K_1 \dfrac{[O_3]}{[O_2]}$

   C.   Rate = $k_2 K_1 \dfrac{[O_3]^2}{[O_2]}$ (combine A. and B.)

Review of Selected Topics

Examples:

Given the overall reaction and the rate equation:
1. Propose a three-step mechanism.
2. Used the mechanism to derive a rate equation in terms of reactant concentrations only. Determine how the overall rate constant, k, relates to the rate constants and equilibrium constants of the mechanism steps.

1. OVERALL REACTION: $I^- + OCl^- \Leftrightarrow OI^- + Cl^-$ in basic solution

   RATE EQUATION: Rate = $k \dfrac{[I^-][OCl^-]}{[OH^-]}$

   Answer:
   MECHANISM:
   1. $OCl^- + H_2O \Leftrightarrow HOCl + OH^-$ (fast equilibrium)
   2. $I^- + HOCl \Rightarrow HOI + Cl^-$ (slow)
   3. $OH^- + HOI \Rightarrow H_2O + OI^-$ (fast)

   A. Rate = $k_2[I^-][HOCl]$ (the slow reaction is step 2)

   B. $K_1 = \dfrac{[HOCl][OH^-]}{[OCl^-]}$ $\quad [HOCl] = K_1 \dfrac{[OCl^-]}{[OH^-]}$

   C. Rate = $k_2 K_1 \dfrac{[O_3][OCl^-]}{[OH^-]}$ $\quad k = K_1 k_2$

2. OVERALL REACTION: $3HNO_2 \Rightarrow H^+ + NO_3^- + 2NO + H_2O(l)$

   RATE EQUATION: Rate = $k \dfrac{[HNO_2]^4}{[NO]^2}$

   Answer:
   MECHANISM:
   1. $4HNO_2 \Leftrightarrow 2NO_2 + 2NO + 2H_2O(l)$ (fast, equil)
   2. $2NO_2 \Leftrightarrow N_2O_4$ (fast, equil)
   3. $N_2O_4 + H_2O(l) \Rightarrow H^+ + NO_3^- + HNO_2$ (slow)

   A. Rate = $k_3[N_2O_4]$ (the slow reaction is step 3)

   B. Liquid water, $H_2O(l)$, is not included in the equilibrium equation for step 2.
   $K_2 = \dfrac{[N_2O_4]}{[NO_2]^2}$ $\quad [N_2O_4] = K_2[NO_2]^2$ (N_2O_4 is an intermediate)

   C. $K_1 = \dfrac{[NO_2]^2[NO]^2}{[HNO_2]^4}$ $\quad [NO_2]^2 = K_1 \dfrac{[HNO_2]^4}{[NO]^2}$ (NO_2 is an intermediate)

   D. Rate = $k_3 [N_2O_4] = k_3 K_2 [NO_2]^2 = k_3 K_2 K_1 \dfrac{[HNO_2]^4}{[NO]^2}$ (Combine A, B and C)

   $k = K_1 K_2 k_3$

Chapter 6
Equilibrium

1. **Equilibrium Constants**

    The equilibrium constant, K, expresses arithmetically the extent to which a reaction will proceed at a given temperature. The equilibrium law is defined in terms of the concentrations or, in the case of gases, the pressures that exist at equilibrium.

    The equilibrium law expression takes two forms for the gaseous equilibrium reaction:
    $$2CO_2(g) \Leftrightarrow 2CO(g) + O_2(g)$$

    1. When concentrations are expressed as molarities:
    $$K_c = \frac{[CO]^2[O_2]}{[CO_2]^2}$$

    2. When concentrations are expressed as partial pressures:
    $$K_p = \frac{(P_{CO})^2(P_{O_2})}{(P_{CO_2})^2}$$

    The equation relating K_c and K_p is:

    $$K_c = K_p \left[\frac{1}{RT}\right]^{\Delta n} \quad \text{where } \Delta n = ((\text{moles of product}) - (\text{moles of reactant}))$$

Example:
A 3:1 starting mixture of hydrogen, H_2, and nitrogen, N_2, comes to equilibrium at 450.°C. The mixture at equilibrium is 9.6% NH_3, 22.6% N_2 and 67.8% H_2 by volume and at 60.0 atmospheres.
What is the value of K_p and K_c for the reaction?

$$N_2(g) + 3H_2(g) \Leftrightarrow 2NH_3(g)$$

Dalton's law of partial pressures predicts that the equilibrium pressures should be:

NH_3: 9.6% x 60.0 = 5.8 atm
N_2: 22.6% x 60.0 = 13.5 atm
H_2: 67.8% x 60.0 = 40.7 atm

$$K_p = \frac{P(NH_3)^2}{P(N_2)P(H_2)^3} = \frac{(5.8)^2}{(13.5)(40.7)^3} = 3.7 \times 10^{-5}$$

$$K_c = K_p \left[\frac{1}{RT}\right]^{\Delta n} = 3.7 \times 10^{-5} \left[\frac{1}{0.0821 \times 723}\right]^{-2} = 0.130$$

47

2. Concentration Changes in Equilibrium Reactions

If the concentration (pressure) of one of the constituents changes at constant temperature, the equilibrium will not be reestablished until all of the equilibrium concentrations change, and the ratio:

$$\frac{[PRODUCTS]}{[REACTANTS]} \quad \text{is again equal to K.}$$

To keep track of the changes occurring as equilibrium is established or re-established, the basic thesis of the 'mole link' procedure used in the 'Stoichiometry' chapter must be modified. It must be taken into account that at equilibrium none of the species is at zero (0) concentration. All of the constituents of the reaction are present to the extent allowed by the numerical value of K. The concentrations changes are followed by using a table that shows the:

- START   (starting concentrations).
- Δ   (delta; the change in concentrations).
- FINISH   (equilibrium concentrations).

Some problems will seem to require that the complete solution will involve the use of the quadratic equation. When the equilibrium constant, K, is less than 0.01 (K<0.01) or greater than 100 (K>100] an important simplification can be made. Unknown values that are added to or subtracted from known values are comparatively small and is ignored.

Examples:

1. Determine the starting pressures for the equilibrium problem in the last section. The percent composition at equilibrium was 9.6% NH_3, 22.6% N_2 and 67.8% H_2 by volume.

   The temperature was at 450.°C and the pressure was 60.0 atmospheres.

   Dalton's law of partial pressures predicts that the equilibrium pressures will be:

   NH_3:   0.096 x 60.0 atm = 5.8 atm
   N_2:   0.226 x 60.0 atm = 13.5 atm
   H_2:   0.678 x 60.0 atm = 40.7 atm

   A key to the solution is to note:

   a. There is no product, NH_3, at the start of the reaction.

   b. The change (5.8 atm) of the ammonia all occurs as equilibrium is established.

   c. During the reaction the stoichiometry of the reaction is followed, and 'mole links' are used extensively.

	$N_2(g)$ +	$3H_2(g)$	⇌	$2NH_3(g)$
Start	16.4 atm	49.4 atm		0.0 atm
Δ	- 2.9	- 8.7		+ 5.8
Finish	13.5	40.7		5.8

2. The reaction $H_2(g) + I_2(g) \Leftrightarrow 2HI(g)$ has an equilibrium constant of 54.4 at 355°C. At 355°C and a total pressure of 0.50 atm, 0.20 mole each of H_2 and I_2 are converted to HI.

   a. What percent of the iodine is converted to product?

	$H_2(g)$ +	$I_2(g)$ $\Leftrightarrow$	$2HI(g)$
Start	0.25 atm	0.25 atm	0.0 atm
Δ	-x	-x	+2x
Finish	0.25-x	0.25-x	2x

$$K_p = \frac{P(HI)^2}{P(H_2) \times P(I_2)} = \frac{(2x)^2}{(0.25-x)^2} = 54.4$$

x = 0.20 atm

% conversion = (0.20 ÷ 0.25) × 100 = **80.%**

   b. What is the numerical value of K_c?

   Since Δn = 0 for this reaction, $K_c = K_p =$ **54.4**

3. The dissociation of H_2S at 900. K has a K_p of 0.0100 atm. What is the percent dissociation at this temperature?

   Let x = dissociated hydrogen sulfide

	$2H_2S(g)$ $\Leftrightarrow$	$2H_2(g)$ +	$S_2(g)$
Start	100. atm	0.0 atm	0.0 atm
Δ	-x	-x	+x/2
Finish	100.-x	x	x/2

The x subtracted from 100, (100-x), is ignored.

$$K_p = 0.0100 = \frac{P_{(PRODUCTS)}}{P_{(REACTANTS)}} = \frac{(x)^2(x/2)}{(100.)^2}$$

x = % dissociation = **5.84%**

The answer would have been 5.6% if x was not ignored.

3. LeChatelier's Principle

The effect of changes in pressure, concentration and temperature on the constituent species and on the equilibrium constant can be predicted. LeChatelier's Principle states that a system in equilibrium will react to a stress in a way that relieves the stress.

CONCENTRATION
If the concentration of a chemical on the reactant side of the reaction is increased, then the reaction to products is favored with a higher rate. The concentration of the reactants is reduced, relieving the stress. This will result in a 'shift' toward increased concentration of the products.

PRESSURE
If the pressure of a system containing gaseous molecules is increased, then the number of molecules must be reduced to relieve the stress. The reaction will 'shift' to the side with the smallest moles of gaseous molecules.

TEMPERATURE
The reaction to products is favored because of a higher forward reaction rate when heat is added to an endothermic reaction (heat is on the reactant side of the reaction). The heat is 'used up' by the endothermic reaction, relieving the stress. The reaction rates is equal after a 'shift' toward higher concentrations of products and a lower concentrations of reactants. The equilibrium constant, K, is larger when heat is added to an endothermic reaction.

When heat is added to an exothermic reaction (heat is on the product side of the reaction), the reaction rate of the products is 'favored'. The result is a decreased product concentration and an increased reactant concentration. The equilibrium constant, K, is smaller.

CATALYSIS
Catalysts do not affect equilibrium concentrations. The addition of the catalyst speeds up the rate of both the forward and reverse reactions. At equilibrium these reaction rates are already equal. The effect of the catalyst is make these equal rates faster.

Example:
Refer to the three gaseous equilibria reactions shown.
All species are gaseous (g).

Reaction A:	N_2O_4	$\Leftrightarrow 2NO_2$	ΔH is positive
Reaction B:	$CO_2 + H_2$	$\Leftrightarrow CO + H_2O$	ΔH is positive
Reaction C:	$SO_2 + \frac{1}{2}O_2$	$\Leftrightarrow SO_3$	ΔH is negative

Fill in the chart by stating the direction of the shift and the effect on the numerical value of K for each change listed in the table. The effects on the reaction mixture are after it has reached its initial equilibrium condition.

The symbols used in the table are:
  '+', '-', and 'none' represent INCREASE, DECREASE, NO EFFECT.
  '0', 'R' and 'L' represent NO EFFECT, SHIFT RIGHT and SHIFT LEFT.

	Reaction A Effect on:		Reaction B Effect on:		Reaction C Effect on:	
	Position	K_p	Position	K_p	Position	K_p
Increase in pressure	Left	0	None	0	Right	0
Increase in temperature	Right	+	Right	+	Left	−
Increase in the partial pressure of XO_2 gas	Left	0	Right	0	Right	0
A catalyst is added	None	0	None	0	None	0
Helium is added. The total pressure is the same.	Right	0	None	0	Left	0

4. Temperature Change and the Arrhenius Equation

The numerical value of the equilibrium constant, K, depends on temperature only, and an increase or a decrease in numerical value follows the predictions of LeChatelier's Principle.

A form of the Arrhenius equation similar to that used in 'Reaction Kinetics' is used to predict a new value of K when temperature is changed.

$$\ln \frac{K_1}{K_2} = -\frac{\Delta H}{R}\left[\frac{1}{T_2} - \frac{1}{T_1}\right]$$

Example:

The reaction:

$$PCl_3(g) + Cl_2(g) \Leftrightarrow PCl_5(g)$$

is exothermic with a heat of reaction, ΔH, of -22.2 kcal mole^{-1}. The value of the equilibrium constant, K, is 0.562 atm^{-1} at 250.°C. What is its value at 200.°C?

$$\ln \frac{K_2}{0.562} = \frac{-(-22,200)}{1.987 \text{ cal mol}^{-1} \text{ K}^{-1}}\left[\frac{1}{473} - \frac{1}{523}\right]$$

$$K_2 = 5.38$$

The equilibrium constant for this exothermic reaction is higher at the lower temperature. The increase in the temperature favors the reverse, endothermic reaction.

5. Solubility Equilibria and K_{sp}

When a 'slightly soluble' or 'insoluble' salt is mixed with water, a saturated solution results and solubility equilibrium is established. The rate of dissociation of the ions from the solid equals the rate of precipitation of the salt.

The equilibrium rules apply to the dissolving of slightly soluble salts to form saturated solutions.

$$\text{Salt(s)} \Rightarrow m \text{ Cation(aq)} + n \text{ Anion(aq)}$$

The unique equilibrium constant, solubility product (K_{sp}) is used to describe this common equilibrium condition. The equation for K_{sp} takes the form:

$$K_{sp} = [\text{cation}]^m[\text{anion}]^n$$

The concentration of all solids is constant and does not appear in the equilibrium law equation. Just as long as there is "SOME", the amount of solid in a saturated solution is not important.

'X' is generally used to denote the molar solubility of a slightly soluble salt. Two commonly found types of problems appear frequently and will be illustrated in the examples. The 'x^2–type' of dissociation is applicable to those salts which break up into two ions (eg AgCl, $BaSO_4$ and $PbCO_3$). Salts which break up into three ions are of the '$4x^3$–type' (eg Ag_2CrO_4, CaF_2 and $Zn(OH)_2$). Recognition of a 'type' will result in a quicker response to solubility questions.

Examples:

1. The solubility of barium chromate, $BaCrO_4$, in water is 1.3×10^{-5} M at 25°C. What is the value of K_{sp} at this temperature?

	$BaCrO_4(s)$	⇔	$Ba^{2+}(aq)$	+	$CrO_4^{2-}(aq)$
Start	Some		0.0 M		0.0 M
Δ	-x		+x		+x
Finish	Some -x		x		x

x is 1.3×10^{-5} M

$K_{sp} = [Ba^{2+}][CrO_4^{2-}] = [x][x] = x^2$

$K_{sp} = [1.3 \times 10^{-5}]^2 = \mathbf{1.7 \times 10^{-10}}$

2. The solubility product of silver chromate, Ag_2CrO_4, in water is 1.2×10^{-12} at 15°C. What is the molar solubility at this temperature?

	$Ag_2CrO_4(s)$ ⇔	$2Ag^+(aq)$ +	$CrO_4^{2-}(aq)$
Start	Some	0.0 M	0.0 M
Δ	-x	+2x	+x
Finish	Some -x	2x	x

$K_{sp} = [Ag^+]^2[CrO_4^{2-}] = [2x]^2[x] = \underline{4x^3} = 1.2 \times 10^{-12}$
x, the molar solubility = 6.7×10^{-5} M

6. **Common Ion Effect**

It is possible to shift the solubility equilibrium to favor the reactants. This is accomplished by adding a solution which contains an ion in common (a 'common ion') with the salt. The result will be a saturated solution which has:

a. a lower solubility of the starting solid.
b. more undissolved solid and a lower ion concentration.

Examples:
What is the solubility of MgF_2?

For MgF_2, the $K_{sp} = 6.4 \times 10^{-9}$ at 27°C.

a. In pure water.

	$MgF_2(s)$ ⇔	$Mg^{2+}(aq)$ +	$2F^-(aq)$
Start	Some	0.0 M	0.0 M
Δ	-x	+x	+2x
Finish	Some -x	x	2x

$K_{sp} = [Mg^{2+}][F^-]^2 = [x][2x]^2 = \underline{4x^3} = 6.4 \times 10^{-9}$
x, the molar solubility = 1.2×10^{-3} M

b. In a 0.10 M of NaF.

	$MgF_2(s)$ ⇔	$Mg^{2+}(aq)$ +	$2F^-(aq)$
Start	Some	0.0 M	0.10 M
Δ	-x	+x	+2x
Finish	Some -x	x	0.10 + 2x

Note: **2x** is small compared to 0.10, and is ignored.

| Finish | Some -x | x | 0.10 |

$K_{sp} = [Mg^{2+}][F^-]^2 = [x][0.10]^2 = 6.4 \times 10^{-9}$
x, the molar solubility = 6.4×10^{-7} M
This is 0.05% of the solubility of MgF_2 in pure water.

c. In a 0.10 M of Mg(NO$_3$)$_2$.

	MgF$_2$(s) ⇔	Mg^{2+}(aq) +	2F$^-$(aq)
Start	Some	0.10 M	0.0 M
Δ	-x	+x	+2x
Finish	Some -x	0.10 + x	2x

Note: x is small compared to 0.10, and is ignored.

Finish	Some -x	0.10	2x

$K_{sp} = [Mg^{2+}][F^-]^2 = [0.10][2x]^2 = 6.4 \times 10^{-9}$
x, the molar solubility = 1.3×10^{-4} M
This is 10% of the solubility of MgF$_2$ in pure water.

7. **Selective Precipitation**

The difference in the solubility of two salts containing a common cation (positive ion) can be used to separate a pair of anions (negative ions). Cations also can be separated by adding a common anion.

Example:
Silver nitrate solution, AgNO$_3$(aq), is added to a mixture which is 0.10 M in sodium chloride, NaCl, and 0.010 M in potassium chromate, K$_2$CrO$_4$. Assume no dilution by the water added along with the AgNO$_3$.

The K$_{sp}$ of AgCl is 1.6×10^{-10} at 25°C.

The K$_{sp}$ of Ag$_2$CrO$_4$ is 9.0×10^{-12} at 25°C.

a. Which precipitates first, AgCl or Ag$_2$CrO$_4$?
What will be the molarity of silver ion, [Ag$^+$], when precipitation begins?

The [Ag$^+$] is calculated for each of the two equilibrium reactions. The precipitation which occurs first is that of the less soluble salt. This lower solubility will be indicated by a smaller equilibrium [Ag$^+$].

The silver chloride, AgCl, equilibrium reaction:

	AgCl(s) ⇔	Ag$^+$(aq) +	Cl$^-$(aq)
Start	Some	0.0 M	0.10 M
Δ	-x	+x	+x
Finish	Some -x	x	0.10 + x

Note: x is small compared to 0.10, and is ignored.

Finish	Some -x	x	0.10

$K_{sp} = [Ag^+][Cl^-] = [x][0.10] = 1.6 \times 10^{-10}$
x, the [Ag$^+$] = 1.6×10^{-9} M

The silver chromate, AgCrO$_4$, equilibrium reaction:

	Ag$_2$CrO$_4$(s) ⇔	2Ag$^+$(aq) +	CrO$_4^{2-}$(aq)
Start	Some	0.0 M	0.010 M
Δ	-x	+x	+x/2
Finish	Some -x	x	0.010 + x/2

Note: x/2 is small compared to 0.10, and is ignored.

Finish	Some -x	x	0.010

K_{sp} = [Ag$^+$]^2[CrO$_4^{2-}$] = [x]^2[0.010] = 9.0 x 10^{-12}
x, the [Ag$^+$] = 3.0 x 10^{-5} M

The silver chloride, AgCl, precipitates out first, at a molarity of silver ion, [Ag$^+$], of 1.6 x 10^{-9} M

b. What is the [Cl$^-$] when the Ag$_2$CrO$_4$ first starts to precipitate?

Part (a) shows that in order to precipitate Ag$_2$CrO$_4$ the [Ag$^+$] must be 3.0 x 10^{-5} M.
K_{sp} = [Ag$^+$][Cl$^-$] = [3.0 x 10^{-5}][Cl$^-$] = 1.6 x 10^{-10}
The [Cl$^-$] = 5.3 x 10^{-6} M
This is 0.005% of the starting chloride ion molarity of 0.10.

8. **Mixtures of Two Solutions and 'Bounce Back'**

The mixing of two solutions may or may not result a precipitate. The outcome of mixing two solutions can be predicted by using the solubility product constant of the solid.

If the product [cation]^m[anion]^n is:

a. greater than K_{sp} the solution exceeds saturation. Precipitation of the solid will occur until the product is equal to K_{sp}.

b. equal to K_{sp} the solution is saturated.

c. less than K_{sp} the solution is unsaturated.
Precipitation will not occur until the product of the ion concentrations exceeds the K_{sp}.

Example:
1. A solution with a final volume of 20. mL is made by mixing 10. mL of 0.10 M Pb(NO$_3$)$_2$ and 10. mL of 0.0010 M Na$_2$SO$_4$. The K_{sp} of PbSO$_4$ is 1.06 x 10^{-8}.
PbSO$_4$(s) ⇔ Pb^{2+}(aq) + SO$_4^{2-}$(aq)
Will a precipitate form?
[Pb^{2+}] = $\frac{10. \text{ mL} \times 0.10 \text{ M}}{20. \text{ mL}}$ = 0.050 M
[SO$_4^{2-}$] = $\frac{10. \text{ mL} \times 0.0010 \text{ M}}{20. \text{ mL}}$ = 0.00050 M
[Pb^{2+}][SO$_4^{2-}$] = [0.050][0.0005] = 2.5 x 10^{-5}
This value exceeds the K_{sp}. Precipitation will occur.

Precipitation reaction calculations make use of the solubility equilibrium expression to find the concentration of ions in the final solution. When two solutions whose product of concentrations exceeds the K_{sp} are mixed, precipitation occurs until solution equilibrium is established. It is possible to simplify these calculations by using the 'bounce back' technique.

It is assumed that during the first change (Δ_1) the system is a reaction that goes to completion, not an equilibrium reaction. The two solutions react until one of the ions is completely used up. Then the system 'bounces back' (Δ_2) to a condition of solubility equilibrium.

Example:

2. Lead iodate, $Pb(IO_3)_2$, is a sparingly soluble salt with a K_{sp} of 2.6×10^{-13} at 25°C. To 35.0 mL of 0.150 M $Pb(NO_3)_2$ solution is added 15.0 mL of 0.800 M KIO_3. A precipitate of $Pb(IO_3)_2$ results.

What are the concentrations of Pb^{2+} and IO_3^- in the final solution? Assume the volumes of the solutions are additive.

$[Pb^{2+}] = \dfrac{35 \text{ mL} \times 0.150 \text{ mol L}^{-1}}{50 \text{ mL}} = 0.150 \text{ M}$

$[IO_3^-] = \dfrac{15 \text{ mL} \times 0.800 \text{ mol L}^{-1}}{50 \text{ mL}} = 0.240 \text{ M}$

Let x = amount of Pb^{2+} formed during 'bounce back'.

	$Pb(IO_3)_2(s)$	⇔ $Pb^{2+}(aq)$	+ $2IO_3^-(aq)$
Start	0	0.105 M	0.240 M
Δ_1	+0.120	-0.105	-0.210
Precip.	Some	0	0.030
Δ_2	-x	+x	+2x
Finish	Some-x	x	0.030 + 2x
Note: 2x is small compared to 0.030, and is ignored.			
Finish	Some-x	x	0.030

$K_{sp} = [Pb^{2+}][IO_3^-]^2 = [x][0.030]^2 = 2.6 \times 10^{-13}$

$[Pb^{2+}] = x = 2.9 \times 10^{-10} \text{ M}$

$[IO_3^-] = 0.030 \text{ M}$

9. Strong Acids and Strong Bases

Strong acids and bases are completely ionized in water solution. $HClO_4$, HI, HBr, HCl and HNO_3 are strong acids, and react with water to give solutions with a hydronium ion, H_3O^+, concentration equal to that of the starting acid. The hydronium ion is conveniently abbreviated as $H^+(aq)$.

1. Example:
A solution is prepared by dissolving 0.10 mole of $HCl(g)$ in 1.0 Liter of water. What is the concentration of ions in the resulting solution?

	$HCl(g)$ +	$H_2O(l)$ $\Rightarrow$	$H^+(aq)$ +	$Cl^-(aq)$
Start	0.10 M	Some	0.0 M	0.0 M
Δ	-0.10	-0.10	+0.10	+0.10
Finish	0.0	Some	0.10	0.10

The Group IA hydroxides (NaOH, KOH, etc.) are strong bases, and are assumed to dissociate completely giving hydroxide ion concentrations equal to that of the starting ionic species.

2. Example:
A solution is prepared by dissolving 0.10 mole of $NaOH(s)$ in 1.0 Liter of water. What is the concentration of ions in the resulting solution?

	$NaOH(s)$ $\Rightarrow$	$Na^+(aq)$ +	$OH^-(aq)$
Start	0.10 M	0.0 M	0.0 M
Δ	-0.10	+0.10	+0.10
Finish	0.0	0.10	0.10

Water ionizes very poorly and the water equilibrium equilibrium constant, K_w, equals 1.0×10^{-14}. The equilibrium law expression for water is: $K_w = [H^+][OH^-] = 1.0 \times 10^{-14}$.

3. Example:
What is the concentration of ions in pure water?
What is the pH of water?

	H_2O	$\Leftrightarrow$	$H^+(aq)$	+	$OH^-(aq)$
Start	Some		0.0 M		0.0 M
Δ	-x		+x		+x
Finish	Some		x		x

$[H^+][OH^-] = x^2 = 1.0 \times 10^{-14}$

At equilibrium:

$x = [H^+] = [OH^-] = 1.0 \times 10^{-7}$

The concentration of $H^+(aq)$ in a solution may be expressed by using the pH scale.

The pH is defined as: $pH = -\log[H^+]$
$pH = -\log[H^+] = 7$

4. Examples:
What is the solution pH in examples 1, 2, and 3?

Example 1: $[H^+]$ = 0.10 M = 1.0×10^{-1} M
             pH = **1.00**

Example 2: $[OH^-]$ = 0.10 M = 1.0×10^{-1} M
             $[H^+][OH^-]$ = 1.0×10^{-14}
             $[H^+]$ = $\dfrac{1.0 \times 10^{-14}}{1.0 \times 10^{-1}}$ = 1.0×10^{-13}
             pH = **13.00**

Example 3: $[H^+]$ = 1.0×10^{-7} M
             pH = **7.00**

5. Example:
What is the pH corresponding to a hydrogen ion concentration, $H^+(aq)$, equal to 5.00×10^{-4}?

The log is conveniently found with a calculator.

$pH = -\log[H^+] = -\log(5.00 \times 10^{-4}) = \mathbf{3.30}$

6. Find the pH of a solution that results from the mixing of 26.0 mL of 0.200 M KOH with 50.00 mL of 0.100 M HBr.
HBr and KOH are strong, and completely ionize.

$[KOH] = [OH^-] = \dfrac{26.0 \text{ mL} \times 0.200 \text{ mol L}^{-1}}{76.0 \text{ mL}} = 0.0684 \text{ M}$

$[HBr] = [H^+] = \dfrac{50.0 \text{ mL} \times 0.200 \text{ mol L}^{-1}}{76.0 \text{ mL}} = 0.0658 \text{ M}$

The 'bounce back' technique is used to get the final concentrations using the water equilibrium reaction.

	H_2O	⇔	H^+(aq)	+	OH^-(aq)
Start	Some		0.0658 M		0.0684 M
Δ_1	+0.0658		-0.0658		-0.0658
Intermed.	Some		0		0.0026
Δ_2	-x		+x		+x
Finish	Some		x		0.0026 + x

Note: x is small compared to 0.0026, and is ignored.

| Finish | Some | | x | | 0.0026 |

$[H^+][OH^-] = [x][0.0026] = 1.0 \times 10^{-14}$
$x = [H^+] = 3.84 \times 10^{-12}$ M
pH = **11.41**

10. Weak Acids, Weak Bases and pH

Most acids and bases are weak. The acid or base reacts with water and is in equilibrium with the resulting ions. The most common definition of acids and bases used in AP Chemistry is the Brönsted-Lowry theory. An acid is a proton–donor (H^+ donor) and a base is a proton–acceptor (H^+ acceptor). Water is 'amphiprotic' and can act as either an acid or a base and its role is determined by the acid or a base added to it.

The extent of an acid–base equilibrium reactions is measured by ionization constants; K_a for acids and K_b for bases. The ionization constants are determined by measuring the pH of the solution or are given as data. K_a and K_b are related by the equation:

$$K_a \times K_b = K_w = 1.0 \times 10^{-14}$$

There are some common elements in the solution of acid-base equilibrium problems.

1. Write the reaction equation.

2. Use start...Δ...finish.

3. If given the pH, H^+ or OH^-:
   Insert values into the table to determine K_a or K_b.

4. If given K_a or K_b:
   Solve for 'x' to determine pH, H^+ or OH^-.

Example:
1. A 0.0500 M solution of acetic acid has a pH of 3.03. What is the value of K_a?

$$CH_3COOH(aq) \Leftrightarrow H^+(aq) + CH_3COO^-(aq)$$

pH = -log [H$^+$] = 3.03
log[H$^+$] = -3.03

Use a calculator to find the inverse log of -3.03.
[H$^+$] = 9.49 x 10^{-4} M

The solution was prepared from acetic acid alone and therefore the number of acetate ions and hydrogen ions are the same.

[CH$_3$COO$^-$] = [H$^+$] = 9.49 x 10^{-4} M

	CH$_3$COOH(aq)	$\Leftrightarrow$	H$^+$(aq)	+	CH$_3$COO$^-$(aq)
Start	0.0500 M		0.0 M		0.0 M
Δ	-9.49 x 10^{-4}		+9.49 x 10^{-4}		+9.49 x 10^{-4}
Finish	0.0491		9.49 x 10^{-4}		9.49 x 10^{-4}

Examine the start...Δ...finish table. There is little difference between the starting and finishing concentrations because x is small compared to 0.0500 M. The usual practice is to use the starting concentration of the acid as the equilibrium amount.

	CH$_3$COOH(aq)	$\Leftrightarrow$	H$^+$(aq)	+	CH$_3$COO$^-$(aq)
Finish	0.0500 M		9.49 x 10^{-4} M		9.49 x 10^{-4} M

$$K_a = \frac{[H^+][CH_3COO^-]}{[CH_3COOH]} = \frac{[9.49 \times 10^{-4}][9.49 \times 10^{-4}]}{[0.0500]}$$

$K_a = 1.80 \times 10^{-5}$

2. (a) What is the K_b for acetate ion, CH_3COO^-?

Acetate ion is the conjugate base of acetic acid.
The K_a for acetic acid is 1.8×10^{-5}.

$K_b = \dfrac{K_w}{K_a} = \dfrac{1.0 \times 10^{-14}}{1.8 \times 10^{-5}} = 5.6 \times 10^{-10}$

(b) What is the pH of a 0.100 M solution of acetate ion?

	CH_3COO^-	+ H_2O	$\Leftrightarrow$ CH_3COOH	+ OH^-
Finish	0.100 M	Some	x	x

$K_b = \dfrac{[CH_3COOH][OH^-]}{[CH_3COO^-]} = \dfrac{x^2}{0.100} = 5.6 \times 10^{-10}$

$x = [CH_3COOH] = [OH^-] = 1.4 \times 10^{-6}$ M

$[H^+] = \dfrac{1.0 \times 10^{-14}}{1.4 \times 10^{-6}} = 1.3 \times 10^{-9}$ M

pH = $-\log[H^+]$ = 8.9

Each proton of a polyprotic acid comes off with more difficulty than the previous one. LeChatelier's Principle predicts the H^+ ions from a first ionization will repress the second ionization. The second ionization <u>can be ignored</u>. The equilibrium is always written for the loss of <u>only one proton</u>.

Example:

3. (a) Calculate the pH of a 0.100 M H_2S solution. The equilibria are:

$H_2S(aq) \Leftrightarrow H^+(aq) + HS^-$ $K_{a_1} = 1.1 \times 10^{-7}$
$HS^-(aq) \Leftrightarrow H^+(aq) + S^{2-}$ $K_{a_2} = 1.2 \times 10^{-13}$

	$H_2S(aq)$	$\Leftrightarrow$ $H^+(aq)$	+ HS^-
Finish	0.100 M	x	x

$K_{a_1} = \dfrac{[H^+][HS^-]}{[H_2S]} = \dfrac{[x][x]}{[0.100]} = 1.1 \times 10^{-7}$

$x = [H^+] = [HS^-] = 1.0 \times 10^{-4}$ M
pH = $-\log[H^+]$ = 4.00

(b) What is the $[S^{2-}]$?

	HS^-	$\Leftrightarrow$ H^+	+ S^{2-}
Finish	1.0×10^{-4} M	1.0×10^{-4}	x

$K_{a_2} = \dfrac{[H^+][S^{2-}]}{[HS^-]} = \dfrac{[1.0 \times 10^{-4}][x]}{[1.0 \times 10^{-4}]} = 1.2 \times 10^{-13}$

$x = [S^{2-}] = 1.2 \times 10^{-13}$ M

11. Buffers

A buffer is an aqueous solution prepared to contain roughly equal amounts of a weak acid and its conjugate base. The buffer resists a change of pH when H$^+$ ions from a strong acid or OH$^-$ ions from a strong base are added. The equilibrium buffer system works by removing the added H$^+$ or OH$^-$ ions by a reaction with an amount of the corresponding base or acid in the system.

The basic equations for buffered solutions for the general reaction, HA $\Leftrightarrow$ H$^+$ + A$^-$, are derived from the equilibrium law expression and are:

$$[H^+] = K_a \times \frac{[HA]}{[A^-]} \qquad\qquad pH = pK_a + \log \frac{[A^-]}{[HA]}$$

The [H$^+$] of a buffer is close to the K_a if the amounts of acid and conjugate base are near to being equal. And the pH of the buffer is close to the pK_a (pK_a = -log [K_a]).

Examples:

1. A buffer solution contains 1.0 M acetic acid ($K_a = 1.8 \times 10^{-5}$) and 1.0 M sodium acetate. What is the pH of the buffer?

$$CH_3COOH(aq) \Leftrightarrow H^+(aq) + CH_3COO^-(aq)$$

[CH$_3$COOH] = 1.0 M   [CH$_3$COO$^-$] = 1.0 M

$$pH = pK_a + \log \frac{[CH_3COO^-]}{[CH_3COOH]} = 4.74 + \log \frac{[1.0]}{[1.0]} = 4.74$$

2. What is the pH of the solution when 0.10 mole of HCl is added to 1.00 L of the buffer prepared in the previous problem?

Use the 'bounce back' technique.

	CH$_3$COOH(aq)	$\Leftrightarrow$ H$^+$(aq)	+ CH$_3$COO$^-$(aq)
Start	1.00 M	0.10 M	1.00 M
Δ_1	+0.10	-0.10	-0.10
Finish$_1$	1.10	0.00	0.90
Δ_2	-x	+x	+x
Finish$_2$	1.10 - x	x	0.90 + x

Note: x is small compared to 1.10 or to 0.90, and is ignored.

Finish$_2$	1.10	x	0.90

$$pH = pK_a + \log \frac{[CH_3COO^-]}{[CH_3COOH]} = 4.74 + \log \frac{[0.90]}{[1.10]} = 4.65$$

A common way to prepare a buffer is to add to a weak acid half as many moles of strong base (like NaOH). The resulting buffer will contain equal moles of acid and conjugate base. The acid equilibrium equation can be used, but the base equation shows the calculation procedure more clearly.

3. A buffered solution is prepared by adding 20.0 mL of a 0.200 M NaOH solution to 50.0 mL of a 0.100 M CH$_3$COOH solution. What is the pH of the final solution?

$$[OH^-] = \frac{20.0 \text{ mL} \times 0.200 \text{ mol L}^{-1}}{70.0 \text{ mL}} = 0.0571 \text{ M}$$

$$[CH_3COOH] = \frac{50.0 \text{ mL} \times 0.100 \text{ mol L}^{-1}}{70.0 \text{ mL}} = 0.0714 \text{ M}$$

	CH$_3$COO$^-$ + H$_2$O	⇔ CH$_3$COOH + OH$^-$		
Start	0	Some	0.0714 M	0.0571 M
Δ$_1$	+0.0571	+0.0571	−0.0571	−0.0571
Finish$_1$	0.0571	Some	0.0143	0.00
Δ$_2$	−x	−x	+x	+x
Finish$_2$	0.0571+x	Some	0.0143+x	x

Note: x is small compared to 0.0571 and 0.0143, and is ignored.

Finish$_2$	0.0571	Some	0.0143	x

$$K_b = \frac{[CH_3COOH][OH^-]}{[CH_3COO^-]} = \frac{(0.0143)x}{0.0571} = 5.6 \times 10^{-10}$$

$$x = [OH^-] = 2.2 \times 10^{-9} \text{ M}$$

$$[H^+] = \frac{1.0 \times 10^{-14}}{2.2 \times 10^{-9}} = 4.5 \times 10^{-6} \text{ M}$$

$$pH = -\log[H^+] = 5.35$$

12. Titration and Neutralization

The concentration of an acid or a base can be determined by volumetric analysis. During the titration procedure a solution of known concentration (the standard solution) is added to a known volume of an solution of unknown concentration.

At the point of neutralization the moles of acid, n$_{acid}$, reacting with the moles of base, n$_{base}$ are determined by the stoichiometry of the reaction equation. If monoprotic acids and bases are used, the moles of acid and base reacting will be the same: n$_{acid}$ = n$_{base}$

Since the moles of solute is equal to the molarity times the volume in Liters, an equation with the same form used with dilutions can be used also for neutralizations.

$$M_{base} V_{base} = M_{acid} V_{acid}$$

Any combination of weak and strong acids or bases can be titrated. The reactions are complete (because H_2O is a product) even though one reactant may be weak and the other strong. The table summarizes results with four possible combinations.

Acid	Base	Resulting solution at the neutralization point is:	Example
Strong	Strong	Neutral	$HCl + NaOH$
Weak	Strong	Basic	$CH_3COOH + NaOH$
Strong	Weak	Acidic	$HCl + NH_3$
Weak	Weak	Beyond scope of syllabus for A.P. Chem	

Examples:

1. A titration is performed to standardize a base. It requires 35.4 mL of the NaOH to neutralize 10.0 mL of 0.104 M HCl. What is the molarity of the base?

   $M_{base}V_{base} = M_{acid}V_{acid}$
   $M_{base} \times (35.4 \text{ mL}) = (0.104 \text{ mol L}^{-1}) \times (10.0 \text{ mL})$
   $M_{base} = 0.0293 \text{ M}$

2. The same titration is performed, except acetic acid, CH_3COOH, is substituted for the HCl. What is the pH of the resulting solution at the equivalence point?

   $n_{CH_3COOH} = n_{NaOH} = (0.0293 \text{ mol L}^{-1}) \times (0.0354 \text{ L})$
   $n_{CH_3COOH} = 0.00104 \text{ moles}$
   $[CH_3COOH] = [OH^-] = \dfrac{0.00104 \text{ mol}}{0.0454 \text{ L}} = 0.0229 \text{ M}$

	CH_3COO^-	+ H_2O	⇔	CH_3COOH	+ OH^-
Start	0	Some		0.0229 M	0.0229 M
Δ_1	+0.0229	+0.0229		-0.0229	-0.0229
Finish$_1$	0.0229	Some		0.00	0.00
Δ_2	-x	-x		+x	+x
Finish$_2$	0.0229+x	Some		x	x

Note: x is small compared to 0.0229, and is ignored.

| Finish$_2$ | 0.0229 | Some | | x | x |

$K_b = \dfrac{[CH_3COOH][OH^-]}{[CH_3COO^-]} = \dfrac{x^2}{0.0229} = 5.6 \times 10^{-10}$

$x = [OH^-] = 1.6 \times 10^{-4} \text{ M}$

$[H^+] = \dfrac{1.0 \times 10^{-14}}{1.6 \times 10^{-4}} = 6.3 \times 10^{-11} \text{ M}$

$pH = -\log[H^+] = \mathbf{10.20}$

Chapter 7
Thermodynamics

1. Calorimetry, Internal Energy and Enthalpy

The 'First Law of Thermodynamics' states that "energy can be converted from one form into another but cannot be created or destroyed". The energy is either in the form of heat, q, or work, w. Heat energy can be converted into work, and work (with systems involving gases) can be converted into heat. The system does not contain heat or work. Heat and work are ways in which energy is transferred to the system.

The transfer of work, w, involves the action of a force that causes (primarily) a change of the volume of a constant pressure system. This is, most importantly, the compression or expansion of a system that contains gases.

Energy transfer as heat, q, occurs when there is a difference in temperature between the system and the surroundings. The heat transfer always occurs from a region of high temperature to a region of low temperature.

Half of the texts use a convention that chemistry students use often, and this convention will be used since it is consistent with other thermodynamic computations. The summary of the convention is shown in the table.

	Work, w	Heat, q
Done on a system	+w	+q (endothermic)
Done by a system	−w	−q (exothermic)

The thermodynamic 'state' functions, internal energy change, ΔE, enthalpy change, ΔH, entropy change, ΔS and free energy change, ΔG, are independent of how a chemical system comes to a 'state'. These "state functions", and their numerical values, are found as data in chemical references. 'Thermo' is concerned with the changes (Δ) in state functions during chemical and physical changes. Thermodynamic data may be used to make predictions about a specified chemical reaction. The predictions possible include:

1. the heat of reaction (ΔH).
2. whether the reaction is spontaneous ($\Delta G < 0$).
3. the value of the equilibrium constant, K.

INTERNAL ENERGY change, ΔE, is a measure of the change of kinetic energy of the system by the application of heat and/or work. Using the convention that heat or work done on the system is absorbed, and heat or work done by the system is evolved:

$\Delta E = q + w$

HEAT, q, is measured in either calories or Joules
(1 calorie = 4.184 Joules).

$q = m_{H_2O} \times \Delta t_{H_2O}$ (in calories)
$q = 4.184 \times m_{H_2O} \times \Delta t_{H_2O}$ (in Joules)

WORK should be converted from pressures and volumes into calories or Joules so it can be added or subtracted from heat.

$w = P\Delta V = \Delta n_{gas}RT$
Use R = 1.987 cal mol^{-1} K^{-1} or 8.314 J mol^{-1} K^{-1}

To convert work in liter atm to Joules multiply by 101:
Work = $\dfrac{R \text{ (in cal)}}{R \text{ (in L atm)}}$ = $\dfrac{1.987 \text{ cal mol}^{-1} \text{ K}^{-1}}{0.0821 \text{ Liter-atm mol}^{-1} \text{ K}^{-1}}$
Work = 24.2 cal Liter^{-1} atm^{-1} × 4.183 J cal^{-1}
Work = **101 J Liter^{-1} atm^{-1}**

1. Example

   A gas absorbs 300. J of heat energy and is compressed from 20.0 L to 10.0 L by an opposing pressure of 2.00 atm.
   What is the ΔE for this process, in Joules?

   $\Delta E = q + w = q + P\Delta V$
   $\Delta E = +300 \text{ J} + (2.00 \text{ atm} \times 10.0 \text{ L} \times 101 \text{ J Liter}^{-1} \text{ atm}^{-1})$
   Answer: $\Delta E = +2330$ Joules

   ENTHALPY CHANGE, ΔH, is a measure of the change of potential energy of a system. It is defined in terms of ΔE.

   $\Delta H = \Delta E + w = \Delta E + (\Delta n)RT$
   Δn is the change in moles of gases during the reaction.
   Δn = (gaseous moles of product – gaseous moles of reactant).

   Example:

2. The ΔE for the sublimation of one mol of iodine is +59.9 kJ at 25°C and 1.00 atm. What is ΔH for the reaction?
   $I_2(s) \Leftrightarrow I_2(g)$

   $\Delta H = \Delta E + (\Delta n)RT$
   $\Delta H = +59.9 \text{ kJ} + ((+1) \times 8.314 \text{ J mol}^{-1} \text{ K}^{-1} \times 298 \text{ K})$
   Answer: $\Delta H = +62.4$ kJ

The internal energy change, ΔE can be determined from the amount of heat evolved in a constant volume, or bomb calorimeter. Most frequently, like in an AP chem experiment, heat is determined in a calorimeter at constant atmospheric pressure. The device, often a covered styrofoam cup, will directly give a measure of enthalpy, ΔH.

$\Delta E = q_v$ \qquad (constant volume bomb calorimeter)

$\Delta H = q_p$ \qquad (constant pressure styrofoam cup calorimeter)

Example:

3. A calibrated bomb calorimeter takes 1500. Joules to raise the temperature of the device and its contents 1.000°C. (The heat capacity of the calorimeter is 1500. J °C^{-1}).

A temperature rise of 3.908°C is measured when the calorimeter is used in the combustion of a 1.000 g sample of powdered terbium, Tb, in excess oxygen.

$$4Tb(s) + 3O_2(g) \Rightarrow 2Tb_2O_3(s)$$

What is the internal energy change, ΔE, for the reaction in kJ mole^{-1} of Tb_2O_3.
Since there is no change in volume:

q_v = 1500. J °C^{-1} × 3.908°C = −5862 J   (Negative sign for the exothermic reaction)

n_{Tb} = $\dfrac{1.000 \text{ g}}{158.9 \text{ g mol}^{-1}}$ = 0.006293 mol

$n_{Tb_2O_3}$ = ½ n_{Tb} = 0.003147 mol

ΔE = $\dfrac{-5562 \text{ J}}{0.003147 \text{ mol}}$ = −1,863,000 J mol^{-1}

Answer: ΔE = −1,863 kJ mol^{-1}

4. A 0.25 mol sample of crystalline ammonium chloride, NH$_4$Cl, is dissolved in 500. mL of pure water in an insulated container at atmospheric pressure and 21.4°C. The solute dissolved with no change in the volume, and the lowest temperature of the solution was 17.8°C. The density and specific heat capacity of the solution is the same as pure water, 1.00 g cm^{-3} and 1.00 cal g^{-1} °C^{-1} respectively.

$$NH_4Cl(s) \Rightarrow NH_4^+(aq) + Cl^-(aq)$$

What is the value of $\Delta H_{solution}$ for NH$_4$Cl (in kcal/mole)?

q = $m_{H_2O} \times \Delta t_{H_2O}$ (in calories)
Since there is no change in pressure:

q_p = 500. g × 3.6°C = +1800 cal   (endothermic reaction)

ΔH = $\dfrac{+1800 \text{ cal}}{0.25 \text{ mol}}$ = +7,200 cal mol^{-1}

Answer: ΔH = +7.2 kcal mol^{-1}

2. Hess' Law and Heat of Reaction

Hess' Law states that the ΔH for any reaction is a constant, which implies that the heats of reaction are additive. The overall heat of reaction is the algebraic sum of the heats of reaction for the reactions that add up to the that of the overall reaction.

Enthalpy changes slightly with temperature and pressure, but changes dramatically with changes in state. In order for a series of reactions to be added, the states under which the heats of reaction are measured must be known. The heat of reaction where all substances are in their 'standard states' at 25°C and 1.0 atmosphere are denoted with a superscript, $\Delta H°$. For example, the standard state for water is the liquid state, not the gaseous or the solid state. Oxygen, O$_2$, is a gas at 25°C and 1.0 atm. and iron, Fe, is a solid.

Example:
1. Find ΔH, in kJ, for the reaction from the standard heat of reaction data given.

$$2XO_2(s) + CO(g) \Rightarrow X_2O_3(s) + CO_2(g)$$

Step	Reaction	ΔH°
1	$XO_2(s) + CO(g) \Rightarrow XO(s) + CO_2(g)$	−26.8 kJ
2	$X_3O_4(s) + CO(g) \Rightarrow 3XO(s) + CO_2(g)$	+ 7.3 kJ
3	$3X_2O_3(s) + CO(g) \Rightarrow 2X_3O_4(s) + CO_2(g)$	−10.6 kJ

a. The overall reaction is used as a guide to manipulate the steps so that they will add up to the overall reaction. A substance on the product side of a equation should cancel equal moles on the reactant side of the next equation.

b. Whatever is done to the reaction is reflected in ΔH°. If the reaction is doubled, ΔH° is doubled. If the reaction is reversed, the sign of ΔH° is changed.

c. When the steps add up to the overall reaction, the ΔH°'s of the steps should equal the overall ΔH°.

For this reaction:
Step 1 is doubled.
Step 2 is reversed, and ⅔ of it is taken.
Step 3 is reversed, and ⅓ of it is taken.

Step	Reaction	ΔH°
1	$2XO_2 + 2CO \Rightarrow \cancel{2XO} + 2CO_2$	−53.6 kJ
2	$\cancel{2XO} + ⅔ CO_2 \Rightarrow \cancel{⅔ X_3O_4} + ⅔ CO$	−4.9 kJ
3	$\cancel{⅔ X_3O_4} + ⅓ CO_2 \Rightarrow X_2O_3 + ⅓ CO$	+ 3.5 kJ

$$2XO_2 + 2CO + CO_2 \Rightarrow X_2O_3 + 2CO_2 + CO$$

The desired overall reaction results when the equation is simplified by combining the reactant and product amounts of CO and CO_2.

$$2XO_2 + CO \Rightarrow X_2O_3 + CO_2$$

Answer: ΔH° = −55.0 kJ

3. Heats of Formation, Combustion and Reaction

The standard molar heat of formation, $\Delta H_f°$, is the enthalpy change for the reaction where 1.0 mole of compound in its standard state is made from its elements in their standard states at 25°C.

$\Delta H_f°$ of C_3H_8 (g)	$3C(s) + 4H_2(g)$	$\Rightarrow C_3H_8(g)$
$\Delta H_f°$ of $Fe_3O_4(s)$	$3Fe(s) + 2O_2(g)$	$\Rightarrow Fe_3O_4(s)$

The standard molar heat of combustion, $\Delta H_c°$, is the enthalpy change for the reaction where 1.0 mole of compound is burned in oxygen to form carbon dioxide and water, with all substances being in their standard states at 25°C.

$\Delta H_c°$ of $CH_3OH(l)$	$CH_3OH(l) + \tfrac{3}{2} O_2(g)$	$\Rightarrow CO_2(g) + 2H_2O(l)$
$\Delta H_c°$ of C_3H_8 (g)	$C_3H_8(g) + 5O_2(g)$	$\Rightarrow 3CO_2(g) + 4H_2O(l)$

The heat of reaction, $\Delta H°$, can be calculated from the heat of formation of the compounds involved in the reaction. The heat of formation of all elements is zero, since the element in such a process would being made from itself.

$$\Delta H° = \Sigma \Delta H_f° \text{ (Products)} - \Sigma \Delta H_f° \text{ (Reactants)}$$

An additional method of calculating heat of reaction would be from standard heat of combustion data. The sum of the heats of combustion in reactions is additive, as is predicted by Hess' Law.

Examples:

1. What is the heat of reaction, $\Delta H°$, in kJ mole^{-1} for the reaction:

$$NH_3(g) + 3F_2(g) \Rightarrow NF_3(g) + 3HF(g)$$

The standard heats of formation, $\Delta H_f°$, are listed in the table under the substance.

	$NH_3(g)$ +	$3F_2(g)$ $\Rightarrow$	$NF_3(g)$ +	$3HF(g)$
$\Delta H_f°$ kJ mol^{-1}	−46.1	0	−125	−271

$\Delta H° = \Sigma \Delta H_f°$ (Products) $- \Sigma \Delta H_f°$ (Reactants)

$\Delta H° = ((1 \times -125) + (3 \times -271)) - ((1 \times -46.1) + (3 \times 0))$

Answer: $\Delta H° = -892$ kJ

2. For the liquids, C_2H_5OH and C_2H_4O, the standard heats of combustion are given in the table.

Find $\Delta H°$, in kcal, for the partial oxidation:

$$2C_2H_5OH(l) + O_2(g) \Rightarrow 2C_2H_4O(l) + 2H_2O(l)$$

	Heat of combustion reactions:	$\Delta H_c°$
1	$C_2H_5OH(l) + 3O_2(g) \Rightarrow 2CO_2(g) + 3H_2O(l)$	–327.6 kcal
2	$C_2H_4O(l) + 5/2 O_2(g) \Rightarrow 2CO_2(g) + 2H_2O(l)$	–279.0 kcal

Reaction 1 should be doubled.
Reaction 2 should be reversed and doubled.

Step	Reaction	$\Delta H°$
1	$2C_2H_5OH(l) + 6O_2(g) \Rightarrow 4CO_2(g) + 6H_2O(l)$	–655.2 kcal
2	$4CO_2(g) + 4H_2O(l) \Rightarrow 2C_2H_4O(l) + 5O_2(g)$	+558.0 kcal

Answer: –97.20 kcal

4. Bond energy

Bond energy is a measure of the strength of covalent bonds. When a bond is broken the potential energy (enthalpy) of the bond is increased. The reaction is endothermic ($\Delta H° > 0$ and $\Delta H°$ is +). When a bond is formed the bond distance is decreased and the reaction is exothermic ($\Delta H° < 0$ and $\Delta H°$ is –).

Bond energies can be used to determine heats of reaction. The algebraic sum of the heat absorbed during the breaking of the bonds of the reactants and the heat released during the formation of products gives the heat of reaction. $\Delta H°$ data may be used to determine bond energy.

Example:
The standard heat of formation of NF_3 is –125 kJ mol^{-1}. The bond energies for N_2 and F_2 are given in the table. Find the average energy for an NF bond, in kJ mol^{-1}.

	$N_2(g)$ +	$3F_2(g)$	$\Rightarrow$	$2NF_3(g)$
Bond energy, kJ mol^{-1}	941	155		'X'

Heat of bond breaking = $(941) + (3 \times 155) = +1406$ kJ
$\Delta H°$ = Bond breaking + Bond formation:
$\Delta H°$ = $(+1406) + (6 \times -'X') = -125$ kJ/mol
Answer: NF bond energy = 'X' = 255 kJ mol^{-1}

5. Entropy

There are two driving forces for chemical reactions. One is enthalpy change, ΔH, which restricts spontaneous reactions to a state of minimum energy. An exothermic reaction, with a ΔH less than zero, is more likely to be spontaneous than an endothermic reaction.

The other driving force is <u>ENTROPY</u>, S. Entropy is a measure of randomness or chaos. The 'second law of thermodynamics' restricts spontaneous reactions to the direction of maximum chaos. Processes with an entropy change, ΔS, greater than zero have increased in chaos, disorder or randomness and are more likely to occur.

The third law of thermodynamics defines minimum entropy and maximum order. The absolute entropy is zero only for pure crystalline solids at absolute zero (0 K). On the basis of the third law absolute entropy values can be calculated.

$$S° = \int c_p \, dT = q_p/T$$

Entropy changes can be calculated from absolute entropies using the procedure similar to that used for enthalpy change.

$$\Delta S° = \Sigma S°_{products} - \Sigma S°_{reactants}$$

Examples:
1. Which would have a ΔS less than zero (ΔS is –)?

    (A) $2H_2O(g)$ $\Rightarrow$ $2H_2(g) + O_2(g)$
    (B) $H_2O(g)$ $\Rightarrow$ $H_2O(l)$
    (C) $CaCO_3(s)$ $\Rightarrow$ $CaO(s) + CO_2(g)$
    (D) $2NH_3(g)$ $\Rightarrow$ $N_2(g) + 3H_2(g)$
    (E)   none of these

Answer: (B)
Liquid water is less chaotic than gaseous water, so the randomness is decreasing in the reaction.

2. Determine the entropy change at 25°C, in $J\,K^{-1}$ for:

	$2SO_2(g)$ +	$O_2(g)$	$\Rightarrow$ $2SO_3(g)$
$S°$ ($J\,K^{-1}\,mole^{-1}$)	248.1	205.03	256.6

$\Delta S°$ = $\Sigma S°_{products}$ – $\Sigma S°_{reactants}$
$\Delta S°$ = (2×256.6) – $((2 \times 248.1) + (1 \times 205.03))$

Answer: $\Delta S° = -188.0 \, J\,K^{-1}$

6. Free Energy

The standard FREE ENERGY change, $\Delta G°$, for a reaction at constant temperature and pressure is defined as:

$$\Delta G° = \Delta H° - \frac{T\Delta S°}{1000}$$

The $T\Delta S°$ term is divided by 1000 because the usual units of entropy are calories or joules, while the units of free energy and enthalpy are kilocalories or kiloJoules.

The free energy change, or net driving force, has no physical significance. It is the solution to an equation which contains the two driving forces for chemical reactions, enthalpy and entropy change. It is possible to determine reaction spontaneity from the addition of the two terms. The result depends on the value and the sign of $\Delta H°$ and $T\Delta S°$.

If $\Delta G°$ is negative, the reaction is spontaneous.
If $\Delta G° = 0$, the reaction is at equilibrium.
If $\Delta G°$ is positive, the reaction is not spontaneous.

The relationship between $\Delta G°$, $\Delta H°$ and $T\Delta S°$.

Process	$\Delta H°$	$T\Delta S°$	$\Delta G° = \Delta H° - T\Delta S°$
1	−	+	−
2	+	−	+
3	+	+	?
4	−	−	?

Process 1 is definitely spontaneous.
    (Both ΔH and ΔS are favorable.)

Process 2 is definitely nonspontaneous.
    (Both ΔH and ΔS are favorable.)

Processes 3 and 4 are possibly (?) spontaneous.
    Process 3 can occur at high temperatures.
        (when $T\Delta S$ is large as well as spontaneous).
    Process 4 can occur at low temperatures.
        (when ΔH is spontaneous and $T\Delta S$ is small and nonspontaneous).

The standard free energy change of formation of a compound, $\Delta G_f°$, is the change in net driving force for a reaction at 25°C. The reactants are elements in their standard states. One (1.0) mole of the compound is formed in its standard state. The free energy of formation, $\Delta G_f°$, of elements is zero, just as it was for the enthalpy of formation, $\Delta H_f°$, of elements. The free energy change of a reaction, $\Delta G°$, is calculated from the free energy of formation of the compounds involved in a reaction using the equation:

$$\Delta G° = \sum \Delta G_f° \text{ (Products)} - \sum \Delta G_f° \text{ (Reactants)}$$

Examples:

1. Estimate the approximate temperature, in °C, at which the reaction given will become spontaneous.

$$CaCO_3(s) \Leftrightarrow CaO(s) + CO_2(g)$$

The reaction is at 25°C, is endothermic and has a positive entropy change. $\Delta H°$ is +42.6 kcal, $\Delta S°$ is +38.4 cal K^{-1} and $\Delta G°$ is + 31.2 kcal. It can be assumed that ΔH and ΔS will not vary in this range.

At the point where the reaction goes from unfavorable to favorable $\Delta G = 0$.

$$\Delta G° = \Delta H° - \frac{T\Delta S°}{1000}$$

$$\frac{T\Delta S°}{1000} = \Delta H°$$

$$T = \frac{(1000) \times (+42.6 \text{ kcal})}{+38.4 \text{ cal K}^{-1}} = 1109 \text{ K} = 836 \text{ °C}$$

Answer: The temperature must be greater than 836°C.

2. Calculate $\Delta G°$ for the reaction:

	4NH$_3$(g) +	5O$_2$(g)	$\Rightarrow$ 4NO(g) +	6H$_2$O(g)
$\Delta G_f°$ kJ mole^{-1}	–16.48	0	86.67	–228.59

$\Delta G° = \Sigma \Delta G_f°$ (Products) $- \Sigma \Delta G_f°$ (Reactants)

$\Delta G° = ((4 \times 86.67) + (6 \times -228.59)) - ((4 \times -16.48) + (0))$

Answer: $\Delta G° = -958.8$ kJ

7. Free Energy and Equilibrium

An important use of free energy change is to determine values of the equilibrium constant, K. ΔG is the free energy change for a reaction containing substances which are <u>not</u> in their standard state. The equation relating ΔG, $\Delta G°$ and the equilibrium constant, K is:

$$\Delta G = \Delta G° + RT \ln K$$

At equilibrium, ΔG = zero (0), and the equation becomes:

$$\Delta G° = -RT \ln K$$

Examples:

1. For the reaction:

$$2CO(g) + O_2(g) \Leftrightarrow 2CO_2(g)$$

The $\Delta G°$ for the reaction is -257.2 kJ mole^{-1}.
What is the equilibrium constant at 25 °C?

$\Delta G°$ = $(-275.2$ kJ$) \times (1000$ J kJ$^{-1})$ = $-275,200$ J

$\Delta G°$ = $-RT \ln K$

$-275,200$ J = $-(8.314$ J mol^{-1} K$^{-1}) \times (298$ K$) \times \ln K$

$\ln K$ = 111.1

Answer: $K_p = 2 \times 10^{48}$
This is K_p since the gaseous state is standard for all of the chemicals in the reaction.

2. $\Delta G° = -9.853$ kcal for the reaction at 298 K:

$$Ag^+(aq) + 2NH_3(aq) \Leftrightarrow Ag(NH_3)_2^+(aq)$$

What is the equilibrium constant, K_c?

$\Delta G°$ = $(-9.853$ kcal$) \times (1000$ cal kcal$^{-1})$ = $-9,853$ cal

$\Delta G°$ = $-RT \ln K$

$-9,853$ cal = $-(1.987$ cal mol^{-1} K$^{-1}) \times (298$ K$) \times \ln K$

$\ln K$ = 16.64

Answer: $K_c = 1.7 \times 10^7$
This is K_c since concentration, as molarities in aqueous solution, is standard for the reaction.

3. Calculate the equilibrium vapor pressure of water at 25°C.

$\Delta G_f° = -56.70$ kcal/mole for $H_2O(l)$

$\Delta G_f° = -54.65$ kcal/mole for $H_2O(g)$

$H_2O(l) \Leftrightarrow H_2O(l)$

K_p = $(P_{H_2O(g)})$ = vapor pressure

$\Delta G°$ = $-RT \ln K_p$

$\Delta G°$ = $(-54.65) - (-56.70)$ = $+2.05$ kcal = $2,050$ cal

$2,050$ cal = $-(1.987$ cal mol^{-1} K$^{-1}) \times (298$ K$) \times \ln K$

$\ln K$ = -3.46

Answer: $K_p = P°_{H_2O} = 0.031$ atm = 24 mm Hg
K_p is in atmospheres. This is a standard unit of pressure.

8. Free Energy Change and Net Cell Potential

The further a oxidation–reduction (redox) reaction is away from equilibrium, the greater will be the voltage or cell potential. The net driving force (free energy change) of a redox reaction is related to cell potential by the equation:

$\Delta G° = -nF\mathcal{E}°$

where:
F = Faradays
F = 23,060 cal mole^{-1} volt^{-1} or 96,500 J mole^{-1} volt^{-1}

$\mathcal{E}°$ = Standard cell potential, volts

Examples:
1. Consider the reaction:

$Mg(s) + 2 H^+(1 \underline{M}) \Leftrightarrow Mg^{2+}(1 \underline{M}) + H_2(1.0 \text{ atm})$

The standard free energy, $\Delta G°$, at 25°C for this reaction is −108.7 kcal mol^{-1}? What would be the standard cell potential, $\mathcal{E}°$, for this reaction at 25°C?

$\Delta G°$ = (−108.7 kcal) × (1000 cal kcal^{-1}) = −108,700 cal
n = 2 moles of electrons are transferred from Mg to 2H$^+$.

$\Delta G° = -nF\mathcal{E}°$

−108,700 cal = − (2.000 mol) × (23,060 cal volt^{-1}) × $\mathcal{E}°$

Answer: $\mathcal{E}° = 2.357$ volts

2. The standard reduction potential, $\mathcal{E}°$, for TeO$_2$ forming Te is +0.593 volt. The standard reduction potential for V^{2+} forming V(s) is −1.175 volt. Calculate the standard free energy change, $\Delta G°$, in kJ, for the reaction:

$V(s) + TeO_2(s) + 4H^+(aq) \Leftrightarrow V^{2+}(aq) + Te(s) + 2 H_2O(l)$

		$\mathcal{E}°$ volts
Oxidation:	$V(s) \Leftrightarrow V^{2+}(aq) + 2e^-$	+1.175
Reduction:	$TeO_2(s) + 4H^+(aq) + 2e^- \Leftrightarrow Te(s) + 2H_2O$	+0.593
Net Cell Potential:		+1.768

$\Delta G° = -nF\mathcal{E}°$
n = 2 moles of electrons are transferred from V to TeO$_2$.
$\Delta G°$ = −(2.00 mol) × (96,500 J volt^{-1}) × (1.768 volts)

Answer: $\Delta G° = -341,000$ J $= -341$ kJ

Chapter 8
Electrochemistry

1. **Balancing Oxidation-Reduction Reactions**

    Reactions in which electrons are transferred are oxidation-reduction or redox reactions. The status of the electron transfer in redox reactions is evaluated by means of oxidation numbers. Every reaction of this type involves at least one element whose oxidation number is increasing and at least one whose oxidation number is decreasing.

REDUCING AGENTS:	OXIDIZING AGENTS:
are oxidized	are reduced
lose electrons	gain electrons
increase in oxidation number	decrease in oxidation number

The rules for assigning oxidation numbers are:

1. The oxidation number of an element is 0.

2. In compounds:

	Oxidation #		Oxidation #
Hydrogen, H	+1	Oxygen, O	-2
Alkali metals	+1	Binary halides	-1
Alkaline earths	+2		
Al and Ga	+3		

3. The sum of the oxidation numbers:

    a. in a compound is 0.

    b. in an ion is the charge on the ion.

The balancing of redox reactions often cannot be accomplished by merely inspecting the number of reactant and product atoms alone. The gain and loss of electrons must also be accounted for in the balancing. An efficient method of balancing redox reactions involves following some rules.

1. Identify the substance being oxidized and the substance being reduced. Expect that only one element in the reaction is oxidized and one element is reduced.
Write an oxidation half–reaction and a reduction half–reaction. Don't worry yet about which is which.

2. Balance each half–reaction with respect to all elements, except hydrogen, H, and oxygen, O.

FOR ACID SOLUTIONS:

3. Add the appropriate number of waters, H$_2$O, to the deficient side to balance the oxygen, O.

4. Add H$^+$ to the deficient side to balance the hydrogens.

FOR BASIC SOLUTIONS:

3. For each deficient oxygen, 'O':
    Add two (2) hydroxides, OH$^-$, to the side that is deficient.
    Add one (1) water, H$_2$O, to the other side.

4. For each deficient oxygen, 'H':
    Add one (1) water, H$_2$O, to the side that is deficient.
    Add one (1) hydroxide, OH$^-$, to the other side.

5. Add the appropriate number of electrons to the side with an excess positive charge of each half–reaction.
    It should be easy to identify which is the oxidation reaction (electrons are on the product side) and which is the reduction reaction (electrons are a reactant).

6. Multiply each half–reaction by integers so the gain of electrons by the reduction reaction equals the loss of electrons by the oxidation reaction.

7. Add the two half–reactions and combine duplications.

Examples:
1. Balance the reaction in acid solution:
    Ag$^+$ + AsH$_3$ $\Rightarrow$ Ag + H$_3$AsO$_3$

Step	Action	Equations
1	HALF–REACTIONS:	Ag$^+$ $\Rightarrow$ Ag AsH$_3$ $\Rightarrow$ H$_3$AsO$_3$
2	BALANCE ATOMS, except O and H:	Ag$^+$ $\Rightarrow$ Ag AsH$_3$ $\Rightarrow$ H$_3$AsO$_3$
3	BALANCE OXYGEN with H$_2$O:	Ag$^+$ $\Rightarrow$ Ag AsH$_3$ + 3H$_2$O $\Rightarrow$ H$_3$AsO$_3$
4	BALANCE HYDROGEN with H$^+$:	Ag$^+$ $\Rightarrow$ Ag AsH$_3$ + 3H$_2$O $\Rightarrow$ H$_3$AsO$_3$ + 6H$^+$
5	BALANCE THE CHARGE WITH ELECTRONS:	Ag$^+$ + e$^-$ $\Rightarrow$ Ag AsH$_3$ + 3H$_2$O $\Rightarrow$ H$_3$AsO$_3$ + 6H$^+$ + 6e$^-$
6	BALANCE THE ELECTRONS:	6Ag$^+$ + 6e$^-$ $\Rightarrow$ 6Ag AsH$_3$ + 3H$_2$O $\Rightarrow$ H$_3$AsO$_3$ + 6H$^+$ + 6e$^-$
7	COMBINE AND SIMPLIFY: 6Ag$^+$ + AsH$_3$ + 3H$_2$O $\Rightarrow$ 6Ag + H$_3$AsO$_3$ + 6H$^+$	

Answer: 6Ag$^+$ + AsH$_3$ + 3H$_2$O $\Rightarrow$ 6Ag + H$_3$AsO$_3$ + 6H$^+$

2. Balance the reaction in basic solution:

$$HS_2O_4^- + CrO_4^{2-} \Rightarrow SO_4^{2-} + Cr(OH)_4^-$$

Step	Action	Equations
1	HALF–REACTIONS:	$HS_2O_4^- \Rightarrow SO_4^{2-}$ $CrO_4^{2-} \Rightarrow Cr(OH)_4^-$
2	BALANCE ATOMS, except O and H:	$HS_2O_4^- \Rightarrow 2SO_4^{2-}$ $CrO_4^{2-} \Rightarrow Cr(OH)_4^-$
3	BALANCE OXYGEN with OH$^-$ and H$_2$O:	$HS_2O_4^- + 8OH^- \Rightarrow 2SO_4^{2-} + 4H_2O$ $CrO_4^{2-} \Rightarrow Cr(OH)_4^-$
4	BALANCE HYDROGEN with H$_2$O and OH$^-$:	$HS_2O_4^- + 9OH^- \Rightarrow 2SO_4^{2-} + 5H_2O$ $CrO_4^{2-} + 4H_2O \Rightarrow Cr(OH)_4^- + 4OH^-$
5	BALANCE THE CHARGE WITH ELECTRONS:	$HS_2O_4^- + 9OH^- \Rightarrow 2SO_4^{2-} + 5H_2O + 6e^-$ $CrO_4^{2-} + 4H_2O + 3e^- \Rightarrow Cr(OH)_4^- + 4OH^-$
6	BALANCE THE ELECTRONS:	$HS_2O_4^- + 9OH^- \Rightarrow 2SO_4^{2-} + 5H_2O + 6e^-$ $2CrO_4^{2-} + 8H_2O + 6e^- \Rightarrow 2Cr(OH)_4^- + 8OH^-$
7	COMBINE AND SIMPLIFY: $HS_2O_4^- + OH^- + 2CrO_4^{2-} + 3H_2O \Rightarrow 2SO_4^{2-} + 2Cr(OH)_4^-$	

Answer: $HS_2O_4^- + OH^- + 2CrO_4^{2-} + 3H_2O \Rightarrow 2SO_4^{2-} + 2Cr(OH)_4^-$

A great deal of information is available from the steps used to balance a redox reaction when the result of each is put into words.

The OXIDATION half–reaction for the reaction is:

$$HS_2O_4^- + 9OH^- \Rightarrow 2SO_4^{2-} + 5H_2O + 6e^-$$

Hydrogen hyposulfite ion, $HS_2O_4^-$, is:

.....oxidized.

.....losing electrons.

.....the reducing agent

.....the oxidation number of each sulfur atom increases from +3 to +6.

The REDUCTION half–reaction for the reaction is:

$$CrO_4^{2-} + 4H_2O + 3e^- \Rightarrow Cr(OH)_4^- + 4OH^-$$

Chromate ion, CrO_4^{2-}, is:

.....reduced.

.....gaining electrons.

.....the oxidizing agent

.....the oxidation number of each chromium atom decreases from +6 to +3.

2. The Faraday and Electrolytic Cells

In an electrolytic cell electrical energy is used to bring about a nonspontaneous electrical change. Electrolysis always results in a redox chemical reaction.

Electrochemical cells, both primary (voltaic) and electrolytic (secondary), consist of an electrolyte and electrodes. An 'electrolyte' is a substance that exists as ions in either the molten (fused) state or in water solution. The 'electrodes' consist of an anode and a cathode which are generally metals or graphite, C, rods. The electrodes conduct electrons to an external circuit which in electrolysis is a source of energy. Electrons <u>always</u> flow from the anode to the cathode.

The anode is <u>always</u> where oxidation (the loss of electrons) occurs.
Remember '<u>AN</u> <u>OX</u>'.
The oxidation reaction is the most spontaneous reaction of the reducing agents.

The cathode is <u>always</u> where reduction (the gain of electrons) occurs.
Remember the '<u>RED</u> <u>CAT</u>'.
The reduction reaction involves the most active of the given oxidizing agents.

Inert electrodes that will not take part in the electrolysis reaction are often selected. Common examples are graphite, C, platinum, Pt, gold, Au, stainless steel or nichrome wire. Anodes which are not inert will oxidize during electrolysis. There are three common electrolytic reactions that occur at the cathode and the anode.

	AT THE CATHODE: (- electrode)	AT AN INERT ANODE: (+ electrode)
1.	A cation (+ ion) is reduced to a metal atom.	An anion (- ion) is oxidized to a nonmetallic molecule.

2.	When the cation is:
	    a Group IA ion or
	    a Group IIA ion
	it <u>will</u> <u>not</u> <u>be</u> <u>reduced</u> in aqueous solution.
	INSTEAD:
	    H_2O is reduced to H_2 + OH^-
		When the anion is:
		    F^-, NO_3^- or SO_4^{2-}
		
		it <u>will</u> <u>not</u> <u>be</u> <u>oxidized</u> in aqueous solution.
		INSTEAD:
		    H_2O is oxidized to O_2 + H^+

3.	With a strong acid:
	
	    H^+ is reduced to H_2
		With a strong base:
		
		    OH^- is oxidized to O_2 + H_2O

80 Review of Selected Topics

Example:

A 1.0 M solution of KI(aq) is electrolyzed between graphite electrodes: Which is true?

(A)   The solution will become more basic.

(B)   Oxygen will be evolved at the anode.

(C)   Iodine will be formed at the cathode.

(D)   Hydrogen will be evolved at the anode.

(E)   Potassium will be deposited at the cathode.

Answer: (A)
KI in an aqueous solution is not a strong acid or base.

At the cathode:
   K^+ is a group IA ion and is not reduced in water solution.
   Thus: H_2O is reduced to $H_2 + OH^-$.

At the anode:
   I^- is oxidized to I_2.

The Faraday is the quantity of electrical charge, in coulombs, that is associated with a mole of electrons. It requires 96,500 coulombs to move one (1) mole of electrons. An ampere is the rate of flow of electrons in coulombs sec^{-1}.

Examples:

1.   A current of 10. amperes is passed through molten magnesium chloride for 3.0 hours. How many moles of magnesium metal could be produced by this electrolysis?

$$n_{electrons} = \frac{10 \text{ coul}}{\text{sec}} \times \frac{1 \text{ mol electrons}}{96,500 \text{ coul}} \times 3.0 \text{ hr} \times \frac{3600 \text{ sec}}{1 \text{ hr}}$$

$n_{electrons}$ = 1.1 moles

At the cathode:
   $Mg^{2+} + 2e^- \Rightarrow Mg$

Using the 'mole link':
n_{Mg} = ½ n_{e^-} = ½ × 1.1 = 0.56 moles

Answer: 0.56 moles

2. One-half (0.500) Faraday of electricity is passed through aqueous solutions of the compounds listed in the table. Electrodes are made of the material indicated.
Give the formula of the substance that is produced at each electrode and the amount in moles that is produced.

Answers:

| | | Cathode || Anode ||
Compound	Electrode material	Substance produced	Amount produced mol	Substance produced	Amount produced mol
Fe(NO$_3$)$_3$	stainless steel	Fe	0.167	O$_2$	0.125

Cathode: $Fe^{3+} + 3e^- \Rightarrow Fe$ $n_{Fe} = 1/3\, n_{e^-}$
Anode: $2H_2O \Rightarrow O_2 + 4H^+ + 4e^-$ $n_{O_2} = 1/4\, n_{e^-}$

| | | Cathode || Anode ||
Compound	Electrode material	Substance produced	Amount produced mol	Substance produced	Amount produced mol
NiSO$_4$	Ni	Ni	0.250	Ni^{2+}	0.250

Cathode: $Ni^{2+} + 2e^- \Rightarrow Ni$ $n_{Ni} = 1/2\, n_{e^-}$
Anode: $Ni \Rightarrow Ni^{2+} + 2e^-$

| | | Cathode || Anode ||
Compound	Electrode material	Substance produced	Amount produced mol	Substance produced	Amount produced mol
Au(NO$_3$)$_3$	Au	Au	0.167	O$_2$	0.125

Cathode: $Au^{3+} + 3e^- \Rightarrow Au$ $n_{Au} = 1/3\, n_{e^-}$
Anode: $2H_2O \Rightarrow O_2 + 4H^+ + 4e^-$ $n_{O_2} = 1/4\, n_{e^-}$

| | | Cathode || Anode ||
Compound	Electrode material	Substance produced	Amount produced mol	Substance produced	Amount produced mol
Li$_2$SO$_4$	Cu	H$_2$	0.250	Cu^{2+}	0.250

Cathode: $H_2O + 2e^- \Rightarrow H_2 + 2OH^-$ $n_{H_2} = 1/2\, n_{e^-}$
Anode: $Cu \Rightarrow Cu^{2+} + 2e^-$ $n_{Cu} = 1/2\, n_{e^-}$

Compound	Electrode material	Cathode Substance produced	Cathode Amount produced mol	Anode Substance produced	Anode Amount produced mol
BaBr$_2$	nichrome	H$_2$	0.250	Br$_2$	0.250

Cathode: $H_2O + 2e^- \Rightarrow H_2 + 2OH^-$ $n_{H_2} = \frac{1}{2} n_{e^-}$
Anode: $Br_2 + 2e^- \Rightarrow 2Br^-$ $n_{Br_2} = \frac{1}{2} n_{e^-}$

Compound	Electrode material	Cathode Substance produced	Cathode Amount produced mol	Anode Substance produced	Anode Amount produced mol
Al$_2$(SO$_4$)$_3$	Pt	H$_2$	0.250	O$_2$	0.125

Cathode: $H_2O + 2e^- \Rightarrow H_2 + 2OH^-$ $n_{H_2} = \frac{1}{2} n_{e^-}$
Anode: $2H_2O \Rightarrow O_2 + 4H^+ + 4e^-$ $n_{O_2} = \frac{1}{4} n_{e^-}$

Compound	Electrode material	Cathode Substance produced	Cathode Amount produced mol	Anode Substance produced	Anode Amount produced mol
SnF$_4$	C	Sn	0.125	O$_2$	0.125

Cathode: $Sn^{4+} + 4e^- \Rightarrow Sn$ $n_{Sn} = \frac{1}{4} n_{e^-}$
Anode: $2H_2O \Rightarrow O_2 + 4H^+ + 4e^-$ $n_{O_2} = \frac{1}{4} n_{e^-}$

Compound	Electrode material	Cathode Substance produced	Cathode Amount produced mol	Anode Substance produced	Anode Amount produced mol
Cd(NO$_3$)$_2$	Cd	Cd	0.250	Cd^{2+}	0.250

Cathode: $Cd^{2+} + 2e^- \Rightarrow Cd$ $n_{Cd} = \frac{1}{2} n_{e^-}$
Anode: $Cd \Rightarrow Cd^{2+} + 2e^-$

Compound	Electrode material	Cathode Substance produced	Cathode Amount produced mol	Anode Substance produced	Anode Amount produced mol
K$_2$SO$_4$	Zn	H$_2$	0.250	Zn^{2+}	0.250

Cathode: $H_2O + 2e^- \Rightarrow H_2 + 2OH^-$ $n_{H_2} = \frac{1}{2} n_{e^-}$
Anode: $Zn \Rightarrow Zn^{2+} + 2e^-$ $n_{Zn} = \frac{1}{2} n_{e^-}$

3. Voltaic Cells

In a voltaic (or primary) cell electrical energy is produced by a spontaneous oxidation–reduction reaction. This spontaneous reaction will occur whenever two dissimilar metallic electrodes are immersed in an electrolyte and connected to an external circuit.

The electrodes of a voltaic cell have charges which are opposite to those of an electrolytic cell. After all, it is working in an opposite manner. The anode is the negative (-) electrode and the cathode is the positive (+) electrode. The 'RED CAT' and 'AN OX' principles stated earlier are true for voltaic cells, as is the rule for the flow of electrons from the anode to the cathode in the external circuit.

An example of a voltaic cell would be the system shown in the diagram. The electrodes are zinc and copper, and the electrolytes in each half–cell include Zn^{2+} and Cu^{2+}. In the solution would also be an anion, such as NO_3^- or SO_4^{2-}, which is a required part of the salt used to prepare the electrolyte.

Zinc is the anode, and the oxidation of Zn(s) to Zn^{2+} occurs at the electrode. The reduction of copper ions, Cu^{2+}, to copper occurs at the other electrode, which is the cathode. The electrons move from the anode to the cathode in the external circuit of every voltaic cell.

A 'salt bridge' containing an nonreactive electrolyte completes the circuit. Potassium chloride solution, KCl(aq), is used in this system. The positive ions in the salt bridge move from the anode to the cathode because reduction is depleting the concentration of positive ions at the cathode. The negative ions move from the cathode to the anode to balance the charge since oxidation is increasing the concentration of positive ions.

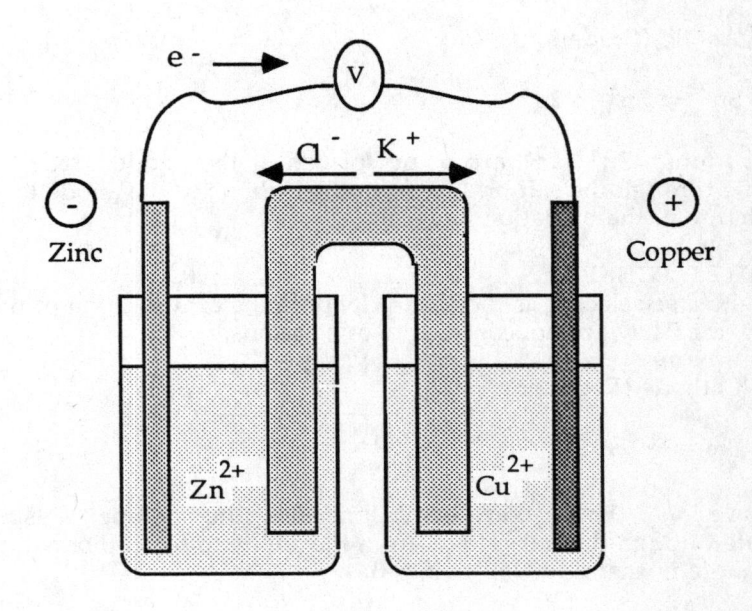

An abbreviation of the cell is:

$$Zn/Zn^{2+} \quad || \quad Cu^{2+}/Cu$$

The three parts of the abbreviation stand for:

⇑ ⇑ ⇑
{anode reaction}   {salt bridge}   {cathode reaction}
$Zn \Rightarrow Zn^{2+} + 2e^-$ $\quad\quad\quad\quad$ $Cu^{2+} + 2e^- \Rightarrow Cu$
{ Oxidation } $\quad\quad\quad\quad\quad\quad$ { Reduction }

Example:
Use the abbreviated form of a voltaic cell to show the arrangement for the reaction:

$$Sn^{2+} + Br_2 \Rightarrow Sn^{4+} + 2Br^-$$

The anode is tin and the cathode is graphite.

Answer:
Sn^{2+}/Sn^{4+} (Sn) || (C) Br_2/Br^-

Electrons flow through the wire connecting the electrodes from the anode, the tin, Sn, electrode (anode) to the cathode, the inert carbon, C, electrode.

NEGATIVE ELECTRODE:
Tin was selected as the electrode. An inert material (Pt or nichrome) could also have been used.

Anode (Oxidation):

$Sn^{2+} \Rightarrow Sn^{4+} + 2e^-$

Tin(IV) ions, Sn^{4+}(aq), are being formed at the anode. Negative ions will always migrate through the salt bridge from the cathode to the anode to maintain a balance of the charge in the oxidation half–cell.

POSITIVE ELECTRODE:
Graphite was selected as the electrode. Bromine or a solution of bromide ion cannot be electrodes. They are nonconductors of electrons.

Cathode (Reduction):

$Br_2 + 2e^- \Rightarrow 2Br^-$

Bromide ions, Br^- (aq), are being formed at the cathode. Positive ions will always migrate through the salt bridge from the anode to the cathode to maintain a balance of the charge in the reduction half–cell.

4. Standard Reduction Potentials

Tables of standard reduction potentials are used to determine cell voltages and to predict whether a redox reaction is spontaneous. The standard state for $\mathcal{E}°$ is the same as it was for the thermodynamic state functions; reactions are at 25°C where gases are at 1.0 atmosphere and solution concentration is 1.0 M.

The more positive the standard reduction potential the stronger is the oxidizing agent and the weaker is its conjugate reducing agent. The $\mathcal{E}°$ for an oxidation half-reaction is the standard reduction potential with the opposite sign.

The table of Standard Electrode Potentials shown below conforms to the style used in the references for Section II of the AP examination. It should be noted that this style does not conform to that used in some texts.

	Reaction			$\mathcal{E}°$	
Weakest	Li^+	$+ e^-$	$\Rightarrow$ Li	- 3.04 volts	Strongest
oxidizing	Zn^{2+}	$+ 2e^-$	$\Rightarrow$ Zn	- 0.76 volts	reducing
agents	Fe^{2+}	$+ 2e^-$	$\Rightarrow$ Fe	- 0.45 volts	agents
	Ni^{2+}	$+ 2e^-$	$\Rightarrow$ Ni	- 0.26 volts	
	Sn^{2+}	$+ 2e^-$	$\Rightarrow$ Sn	- 0.14 volts	
	$2H^+$	$+ 2e^-$	$\Rightarrow$ H_2	0.00 volts	
	Sn^{4+}	$+ 2e^-$	$\Rightarrow$ Sn^{2+}	0.15 volts	
	Cu^{2+}	$+ 2e^-$	$\Rightarrow$ Cu	0.34 volts	
	I_2	$+ 2e^-$	$\Rightarrow$ $2I^-$	0.54 volts	
	Fe^{3+}	$+ 2e^-$	$\Rightarrow$ Fe^{2+}	0.77 volts	
	Ag^+	$+ e^-$	$\Rightarrow$ Ag	0.80 volts	
Strongest	Br_2	$+ 2e^-$	$\Rightarrow$ $2Br^-$	1.09 volts	Weakest
oxidizing	Au^{3+}	$+ 3e^-$	$\Rightarrow$ Au	1.50 volts	reducing
agents	F_2	$+ 2e^-$	$\Rightarrow$ $2F^-$	2.87 volts	agents

The cell voltage is determined by adding the potentials of the oxidation and reduction half-reactions. The half-reaction potential <u>does not change</u> when the reaction is multiplied to balance the electrons.

If the cell voltage is positive, the reaction will occur spontaneously. If the voltaic cell has a negative potential, the reverse reaction will be spontaneous. When $\mathcal{E}°$ is 0, the reaction will be at equilibrium in the standard state.

Examples:
1. What is the net cell potential and the overall reaction for the voltaic cell?
   The abbreviated form is: Ni/Ni^{2+} || Ag$^+$/Ag

   Anode (Oxidation):

   $$Ni \Rightarrow Ni^{2+} + 2e^- \qquad \mathcal{E}° = +0.26 \text{ volts}$$

   For an oxidation reaction, the sign of the reduction potential is changed.

   Cathode (Reduction):

   $$2Ag^+ + 2e^- \Rightarrow 2Ag \qquad \mathcal{E}° = +0.80 \text{ volts}$$

   The cathode reaction is doubled to balance the electrons.
   Answer:

   $$Ni + 2Ag^+ \Rightarrow Ni^{2+} + 2Ag \qquad \mathcal{E}° = +1.06 \text{ volts}$$

2. Which will be oxidized by Ni^{2+}?
   (A) Ag
   (B) Br$^-$
   (C) Cu
   (D) F$_2$
   (E) Fe

   Answer (E) Fe has an oxidation potential (+0.45 volts) high enough to give a positive overall potential.

   $$Fe \Rightarrow Fe^{2+} + 2e^- \qquad \mathcal{E}° = +0.45 \text{ volts}$$

   $$Ni^{2+} + 2e^- \Rightarrow Ni \qquad \mathcal{E}° = -0.26 \text{ volts}$$

   $$Fe + Ni^{2+} \Rightarrow Fe^{2+} + Ni \qquad \mathcal{E}° = +0.19 \text{ volts}$$

3. Which will be reduced by Ag?
   (A) Au^{3+}
   (B) Cu
   (C) Fe^{3+}
   (D) I$_2$
   (E) Ni^{2+}

   Answer: (A) Au^{3+} is the only choice with a high enough reduction potential (+1.50 volts) to react with silver, Ag, to give a spontaneous overall reaction.

   $$3Ag + Au^{3+} \Rightarrow 3Ag^+ + Au \qquad \mathcal{E}° = +0.70 \text{ volts}$$

5. The Nernst Equation

An electrochemical cell is a system in the process of achieving chemical equilibrium. The highest voltage occurs when the system is far from equilibrium. LeChatelier's Principle is used to analyze the effect of stresses which are applied or removed. Cell potential is independent of the size of the electrodes, the size of the cells and the volume of electrolyte since the constant concentration of solids and liquids does not affect the kinetics of the equilibrium process. Concentration does affect the voltage. Higher reactant and lower product concentration will increase cell potential.

The Nernst equation is used to determine the voltage of a voltaic cell when the concentrations are not 1.0 M.

(Nernst Equation) at 25°C; $\quad \mathcal{E} = \mathcal{E}° - \dfrac{0.0592}{n} \log K$

Example:

1. The electrolytes of the cell are 0.0500 M $Zn(NO_3)_2$ and 1.000 M $Cu(NO_3)_2$. What will be the initial cell voltage?

$Zn/Zn^{2+} \; || \; Cu^{2+}/Cu$

$Zn + Cu^{2+} \Rightarrow Zn^{2+} + Cu$ $\qquad\qquad \mathcal{E}° = 1.10$ volts

$\mathcal{E} = \mathcal{E}° - \dfrac{0.0592}{n} \log K = \mathcal{E}° - \dfrac{0.0592}{n} \log \dfrac{[Zn^{2+}]}{[Cu^{2+}]}$

n = 2 moles of electrons transferred

$\mathcal{E} = 1.10 - \dfrac{0.0592}{2} \log \dfrac{[0.0500 \text{ M}]}{[1.000 \text{ M}]} = 1.10 - (-0.04) = 1.14$ volts

Answer: $\mathcal{E} = 1.14$ volts

2. Calculate the voltage, $\mathcal{E}$, for the same cell when the cell has discharged to the point where the $[Cu^{2+}]$ is 0.010 M.

Assume that the zinc electrode is not entirely consumed.

	Zn	+	Cu^{2+}	$\Rightarrow$	Zn^{2+}	+	Cu
Start	—		1.000		0.0500		—
Δ	−0.990		−0.990		+0.990		+0.990
Finish	Some		0.010		1.040		Some

$\mathcal{E} = 1.10 - \dfrac{0.0592}{2} \log \dfrac{[1.040]}{[0.010]} = 1.10 - (+0.06) = 1.04$ volts

Answer: $\mathcal{E} = 1.04$ volts

Part II

Multiple Choice Questions

About Section I

The Multiple Choice Questions

Section I of the AP Chemistry exam constitutes 45% of the grade, and consists of 75–85 multiple choice questions. Usually 75 questions are asked and ninety (90) minutes is allowed. The 85 question test allows 115 minutes for the section and includes chemical equations.

Chemical equations are reviewed where they are commonly tested—in Section II.

Only about one (1) minute should be spent on each question. There are questions at the end of the section that will test concepts that are familiar, so it is important to get to them. Here are some additional hints.

1. The scoring formula is:

    Total Correct = #Right - (#Wrong ÷ 4).

    Answer only those questions where the right answer is known, or some of the choices can be eliminated. If 1 out of 5 items involving guesses is answered correctly, it's a wash. But 2 or more out of 5 answered correctly represents a significant gain.

2. Don't change an answer unless an obvious error is detected. Nine times out of 10 the first choice is the best one.

3. A time-tested axiom is "If you think long, you think wrong". The 'distractors' (wrong answers) are intentionally distracting.

4. Read the question carefully. Write on the exam paper. Underline the key chemical terms. Pay attention to dimensional units and significant figures.

5. Eliminate answers that are internally inconsistent. An exothermic reaction can <u>never</u> absorb energy. Sodium salts will not precipitate.

6. Some of the questions <u>are</u> surprisingly easy. Take the point (and the time) to move on to more challenging questions.

7. Depend on careful preparation and reading only! <u>Forget</u> the rule "When in doubt, pick answer (C) or (D)". The test is written by professionals. The key is most likely statistically checked. A string of two or three (A) answers is possible statistically, and so is any other trend.

Set 1

Atomic Structure and Periodicity

Directions for Questions 1–3:
Select an answer from the five atomic numbers given. You may use an atomic number once, more than once, or not at all.

(A) 4
(B) 8
(C) 9
(D) 17
(E) 85

1. This element has the highest ionization energy of the elements listed.

2. All isotopes of this element are naturally radioactive.

3. This element has two s– and only four p– electrons in its valence shell.

4. Which mass–spectrometer graph represents naturally occurring boron?

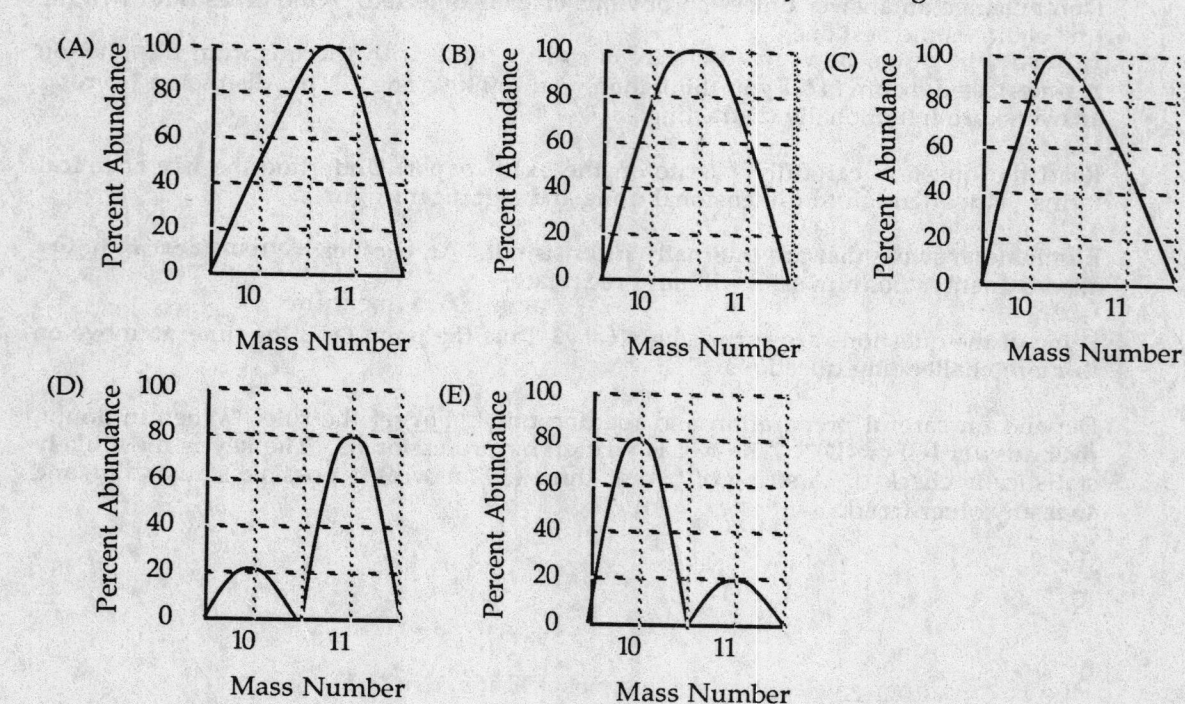

Atomic Structure and Periodicity

5. Two isotopes of silicon are Si–27 and Si–31. Both would be expected to have the same
   - (A) half–life
   - (B) nuclear charge
   - (C) atomic mass
   - (D) number of neutrons
   - (E) decay particle

6. Which of the electron transitions shown in the diagram would evolve the most energy?

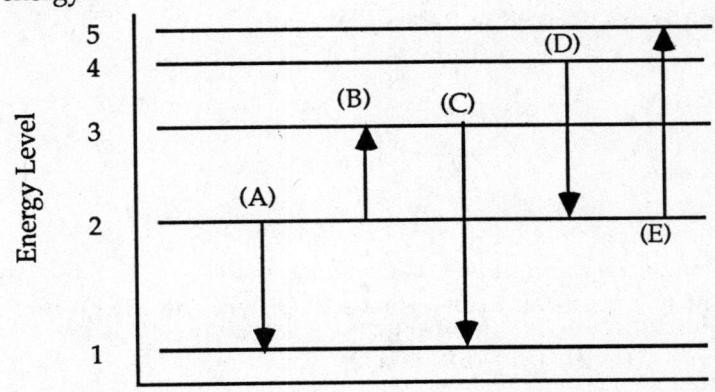

7. In a specific atom, how many electrons can there be with the quantum designation $n = 3$ and $l = 2$?
   - (A) 14
   - (B) 10
   - (C) 8
   - (D) 6
   - (E) 2

8. The assignment of the quantum number $l = 0$ to a 1s– electron implies that the electron has no:
   - (A) energy
   - (B) orbital
   - (C) magnetic field
   - (D) angular momentum
   - (E) spin

9. Ions having the electron structure
   $1s^2\ 2s^2 2p^6$
   would be present in an aqueous solution of:
   - (A) LiBr
   - (B) CaBr$_2$
   - (C) NaF
   - (D) KCl
   - (E) SrI$_2$

10. How many <u>unpaired</u> electrons are in a nickel III ion, (Ni^{3+})?
    - (A) 1
    - (B) 2
    - (C) 3
    - (D) 4
    - (E) 5

11. A magnesium atom (Mg°) differs from a magnesium ion (Mg^{2+}) in that the <u>atom</u> has:
    (A) 3s electrons.
    (B) a higher ionization energy.
    (C) a greater stability.
    (D) a smaller radius.
    (E) fewer subshells.

12. Which would exhibit paramagnetism in the gaseous state?
    I   Al   IV Mg
    II Ar   V   Na
    III Hg

    (A) I, II and III only   (D) I, III and V only
    (B) I and V only   (E) Only II
    (C) II and IV only

13. The transition element which has all its *d*– orbitals filled when in the ground state is:
    (A) Ag   (D) Sn
    (B) Cr   (E) Ga
    (C) Fe

14. Which element would form colored cations and have multiple oxidation states?
    (A) Sc   (D) Zn
    (B) Mn   (E) Ag
    (C) F

15. Which set of elements is listed in order of increasing ionization energy?
    (A) Sb < As < P < S < Cl   (D) P < As < Sb < S < Cl
    (B) Cl < Sb < P < As < S   (E) Sb < As < Cl < S < P
    (C) As < Cl < P < S < Sb

16. Lithium (Li) has the highest ionization energy of the alkali metals (Group IA). But Li$^+$ has the lowest reduction potential of the group. Which is an explanation of this discrepancy?
    (A) Lithium ions have a high hydration energy.
    (B) Ionization energy is endothermic and reduction potential is exothermic.
    (C) Solid lithium has a strong metallic bond.
    (D) Less energy is required to remove an electron when the reaction is carried out in solution.
    (E) Ionization energy of metals and reduction potential of metal ions are measurements of exactly the same reaction, but going in different directions.

17. Why does gold have a higher first ionization energy than silver?
    (A) Gold is a noble metal, silver is not.
    (B) Gold causes the halogens to gain electrons more easily than silver.
    (C) Gold atoms are the same size as silver atoms, but have a higher nuclear charge.
    (D) Gold dissolves in a HCl–HNO$_3$ mixture called 'aqua regia', silver only dissolves in HNO$_3$.
    (E) The screening effect of f– subshell electrons is present in silver, but absent in gold.

18. Electron affinity generally increases as the atomic number increases in Period 2. However, the electron affinity <u>decreases</u> between
    I $_3$Li and $_4$Be   IV $_6$C and $_7$N
    II $_4$Be and $_5$B   V $_7$N and $_8$O
    III $_5$B and $_6$C

    (A) I only
    (B) I and IV only
    (C) II and V only
    (D) II and IV only
    (E) V only

19. In the decay series that starts with uranium–238, the first four steps consist of the emission of an alpha, two beta and then another alpha particle. What is the daughter isotope after the fourth emission?

    (A) $^{234}_{92}$U (B) $^{230}_{90}$Th (C) $^{234}_{90}$Th (D) $^{230}_{88}$Ra (E) $^{230}_{86}$Rn

20. Bismuth–210 has a half–life of 5.0 days. Approximately how many days would it take for 87.5% of a 1.00 mg sample of this isotope to beta–decay to polonium–210?
    (A) 1 day
    (B) 4.4 days
    (C) 5.8 days
    (D) 9.4 days
    (E) 15 days

Answer Key—Set 1

1.	(C)	6.	(C)	11.	(A)	16.	(A)
2.	(E)	7.	(B)	12.	(B)	17.	(C)
3.	(B)	8.	(D)	13.	(A)	18.	(B)
4.	(D)	9.	(C)	14.	(B)	19.	(B)
5.	(B)	10.	(C)	15.	(A)	20.	(E)

Explanations

1. **(C)** The highest ionization energies for elements are found in the top right corner of the periodic table. Only helium (He, element 2) and neon (Ne, element 10) have a higher first ionization energy than fluorine (F, element 9,).

2. **(E)** All isotopes of elements whose atomic number is higher than 83 are naturally radioactive.

3. **(B)** The element has six valence electrons, is in the fourth box of the *p*– block. It will be a member of the oxygen family of elements.

4. **(D)** Boron is found as two naturally occurring stable isotopes, boron–10 and boron–11. The atomic weight is a weighted average, and its value of 10.8 suggests there is 80% boron–11 and 20% boron–10.

5. **(B)** Since both isotopes are silicon, they would have to have the same atomic number and hence the same number of protons in the nucleus.

6. **(C)** Energy is evolved as the electron moves from higher to lower energy shells. The energy varies inversely with n^2, making the energy differences between shells greater at lower energies.

7. **(B)** The letter designation is the *3d*– subshell, which has 5 orbitals that can accomodate 10 electrons.

8. **(D)** The 1s orbital is spherical in shape, and has no angular momentum.

9. **(C)** Sodium ions have 10 electrons. Ions generally have the same number of electrons as the nearest noble gas.

10. **(C)** Nickel has the electron structure $[Ar]3d^8 4s^2$. When three electrons are lost, the structure becomes $[Ar]3d^7$. Three of the five *d*– orbitals have unpaired electrons.

11. **(A)** Magnesium has the electron structure $[Ne]3s^2$. The 3s–electrons are lost when a magnesium ion is formed.
    The magnesium atom, Mg, is larger and has more subshells than the ion. The atom also has a lower ionization energy and is therefore less stable than the ion (Mg^{2+} has a noble gas electron structure).

12. **(B)** Paramagnetism is caused by unpaired electrons. Aluminum and sodium have an odd number of electrons and one unpaired electron each.

13. **(A)** Silver has the electron structure [Kr]4d^{10}5s^1. Chromium and iron have incomplete $d-$ subshells. Tin and gallium are not transition elements.

14. **(B)** Transition elements with incomplete $d-$ subshells generally exhibit multiple oxidation states and have colored ions.

15. **(A)** The highest ionization energies occur in atoms at the top right corner of the periodic element. The lowest ionization energies occur in atoms at the bottom left hand corner of the periodic element.

16. **(A)** Reduction potentials are measured in 1.0 M water solutions. When an electron is removed from an atom to form an ion, energy is absorbed. When the ion hydrates (bond formation) energy is evolved. The large hydration energy associated with the small lithium ion lowers its reduction potential.

17. **(C)** Gold is the same approximate size as silver because of what is called the 'lanthanide contraction'. The *4f*-subshell lowers the shielding of the valence electrons from that which would be expected.

18. **(B)** Electron affinity is the amount of energy released when an electron is added to a neutral atom. Less energy is evolved when an electron is added to either an exactly filled or exactly half-filled sublevel, since these arrangements are stable electron-configurations.

19. **(B)** Two alphas reduce the mass of the daughter by 8 and the atomic number by 4. Two betas increase the atomic number of the daughter by 2. The net is to decrease the mass by 8 and the atomic number by 2.

20. **(E)** $\ln \dfrac{N_o}{N_t} = -\dfrac{0.693\, t}{t_{1/2}}$

    After time t, 87.5% (0.875) has decayed and 12.5% (0.125) remains.

    N_t/N_o is the fraction <u>remaining</u>.

    $\ln 0.125 = -\dfrac{0.693\, t}{5.0 \text{ days}}$

    $t = \dfrac{(-2.08)(5.0 \text{ days})}{(-0.693)} = \mathbf{15.0}$ **days**

Set 2

Chemical Bond

For questions 1-4 select an answer from the five bond types given. You may use a bond type once, more than once, or not at all.

(A) Coordinate covalent bond
(B) Covalent bond
(C) Hydrogen bond
(D) Ionic bond
(E) Metallic bond

1. Which is the type of bond responsible for the linkage between atoms having a similar electron affinity and a high electronegativity?

2. Which is the bond which best accounts for the high melting points and low electrical conductivity of crystalline alkali metal salts?

3. Substances with this bonding have excellent thermal and electrical conductivity.

4. Transitional metal ions form d^2sp^3 octahedral complexes using this bond-type to link with ligands.

5. Which is always associated with chemical bonding?
   (A) Eight electrons in the valence shell of each atom in the compound.
   (B) Two electrons in the valence shell of each atom in the compound.
   (C) Filled valence shells in the atoms to be bonded.
   (D) The attractive forces in the compound are less than the attractive forces of the component atoms.
   (E) The attractive forces in the compound are greater than the attractive forces of the component atoms.

6. Which compound could be formed from the hypothetical elements M and Z? They have the Lewis electron-dot structures to the right.

   (A) M:Z:
   (B) M:Z:
   (C) M:Z:
   (D) $M^{2+}[:Z:]^{1-}_2$
   (E) $M^{1+}_2[:Z:]^{2-}$

98

7. Which is the valid Lewis structure for the sulfite ion (SO$_3^{2-}$)?

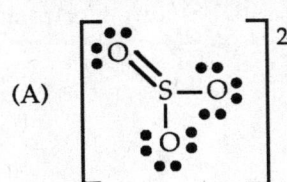

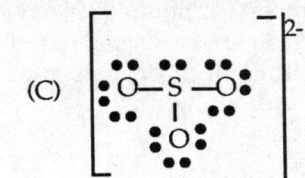

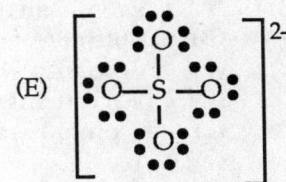

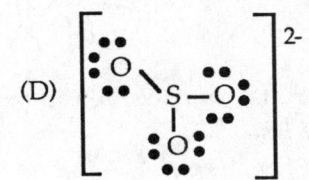

8. Which would have a resonance structure?

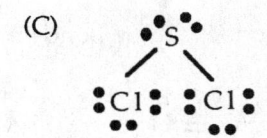

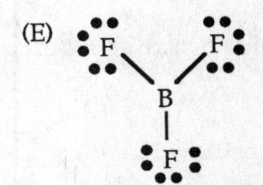

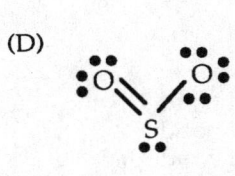

9. What is the geometric shape of the ClF$_4^+$ ion?
   (A) irregular pyramid (see–saw)
   (B) octahedron
   (C) square planar
   (D) tetrahedral
   (E) trigonal pyramidal

10. XF$_3$ has a trigonal pyramidal molecular structure. To which main group of the Periodic Table does 'X' belong?
    (A) III (B) IV (C) V (D) VI (E) VII

11. What is the geometry of BrF$_3$?
    (A) octahedron
    (B) planar triangle
    (C) trigonal bipyramidal
    (D) trigonal pyramidal
    (E) T–shaped

12. In the PCl$_6^-$ ion, what is the Cl–P–Cl bond angle?
    (A) 60° (B) 90° (C) 109.5° (D) 120° (E) 180°

13. Molecular orbital theory predicts that Ne₂ will not form bonds because:
    (A) Neon has equal numbers of bonding electrons and anti–bonding electrons.
    (B) Molecules with electrons in σ* orbitals are unstable.
    (C) Molecules with electrons in π* orbitals are unstable.
    (D) Ne₂ can only form weak π_{2p} bonds.
    (E) Neon is paramagnetic.

14. According to molecular orbital theory, which has the highest bond energy and the smallest bond distance?
    (A) Be₂   (B) NO   (C) NO⁺   (D) NO⁻   (E) O₂

15. Which is the molecular orbital structure for an oxygen molecule, O₂?
    (A) $(\sigma_{2s}^b)^2 (\sigma_{2s}^*)^2 (\pi_{2p}^b)^4 (\pi_{2p}^*)^4$
    (B) $(\sigma_{2s}^b)^2 (\sigma_{2s}^*)^2 (\sigma_{2p}^b)^2 (\sigma_{2p}^*)^2 (\pi_{2p}^b)^4$
    (C) $(\sigma_{2s}^b)^2 (\sigma_{2s}^*)^2 (\sigma_{2p}^b)^2 (\pi_{2p}^b)^3 (\pi_{2p}^*)^3$
    (D) $(\sigma_{2s}^b)^2 (\sigma_{2s}^*)^2 (\sigma_{2p}^b)^2 (\pi_{2p}^b)^4 (\pi_{2p}^*)^2$
    (E) $(\sigma_{2s}^b)^2 (\sigma_{2s}^*)^2 (\sigma_{2p}^b)^2 (\pi_{2p}^b)^4 (\pi_{2p}^*)^4 (\sigma_{2p}^*)^2$

16. Which description best describes all the bonds in acrylonitrile, CH₂CHCN?

    (A) 9 pi bonds
    (B) 9 sigma bonds
    (C) 4 sigma bonds and 5 pi bonds
    (D) 5 sigma bonds and 4 pi bonds
    (E) 6 sigma bonds and 3 pi bonds

17. Which molecule has sp² hybridization of a carbon atom?
    (A) CH₂CH₂
    (B) CH₃CH₃
    (C) CHCH
    (D) CH₄
    (E) CO₂

18. What is the correct name of K₂[PtCl₄]?
    (A) potassium tetrachloroplatinum
    (B) potassium tetrachloroplatinum II
    (C) potassium tetrachloroplatinum IV
    (D) potassium tetrachloroplatinate II
    (E) potassium tetrachloroplatinate IV

19. What is the number of ions in the coordination compound, [Ir(NH₃)₃Cl₃]Cl₃?
    (A) two   (B) four   (C) six   (D) seven   (E) ten

20. Which could not be a ligand?
    (A) H₂O
    (B) CO
    (C) CN⁻
    (D) NH₂CH₂=CH₂NH₂
    (E) CH₄

Chemical Bond   101

Answer Key—Set 2

1.	(B)	6.	(D)	11.	(E)	16.	(E)
2.	(D)	7.	(C)	12.	(B)	17.	(A)
3.	(E)	8.	(D)	13.	(A)	18.	(D)
4.	(A)	9.	(A)	14.	(C)	19.	(B)
5.	(E)	10.	(C)	15.	(D)	20.	(E)

Explanations

1. **(B)** Covalent bonds occur between nonmetals, which are characterized by high electron affinities and high ionization energies.

2. **(D)** The alkali metal salts are made up of cations and anions electrostatically attracted to each other forming a strong crystalline structure. Since the ions cannot move, the salt will not conduct electricity in the solid state.

3. **(E)** Metals are characterized by delocalized electrons. These mobile electrons account for the high thermal and electrical conductivity.

4. **(A)** The transition metal ions are electron–deficient, which allows the acceptance by the central atom of nonbonded (lone–pair) electrons from ligands.

5. **(E)** The attractive forces in compounds must be greater than in the component atoms or the bond would not occur. Most, but not all atoms bond to form stable octets. Only hydrogen has two electrons in the valence shell after bonding. Filled valence shells do not tend to bond (e.g. noble gases).

6. **(D)** Each 'M' will donate or share 2 electrons. Each 'Z' will accept one electron or share one electron.

7. **(C)** Count the electrons. The sulfur and the three oxygens have a total of 24 valence electrons. The 2⁻ charge on the ion accounts for two more, for a total of 26. Three bonds use 6, leaving 20 electrons (10 pairs) to complete octets around each atom.

8. **(D)** Resonance occurs in structures with single <u>and</u> double bonds. Equivalent Lewis structures for SO_2 can be written by switching the single and double bonds.

9. **(A)** ClF_4^+ has 10 electrons surrounding the chlorine and a coordination number of 5. The lone nonbonding pair will be on the plane of the trigonal bipyramid.

10. **(C)** The Group V elements have 5 valence electrons. The result in a structure with a lone pair and 3 unpaired (bonding) electrons.

11. **(E)** BrF_3 has 10 electrons surrounding the bromine and a coordination number of 5. Two nonbonding pairs will be on the plane of the trigonal pyramid, resulting in a T-shaped bonding structure.

12. **(B)** The PCl_6^- ion has 12 bonding electrons or 6 bonds between the chlorines and the central phosphorus atom, resulting in an octahedral shape.

13. (A) Ne₂, with a total of 16 valence electrons, would have 8 in bonding orbitals and 8 in antibonding orbitals. The bond order is zero.

14. (C) NO⁺ has a bond order of 6. NO has a bond order of 5. O₂ and NO⁻ each have a bond order of 4. Be₂ has a bond order of zero. The high bond orders exhibit high bond energy and small bond lengths.

15. (D) Lower energy bonding orbitals are filled before higher energy antibonding orbitals.

16. (E) Single bonds are sigma bonds. In a multiple bond, the first bond is a sigma bond and subsequent bonds are pi bonds.

17. (A) sp² hybridization occurs when the carbon atom forms three bonds. A double or triple covalent bond counts as one bond.

18. (D) When the complex ion is the anion, the ending -ate is used. The oxidation number of the two potassium ions totals +2, and that of the four chloro– ligands is -4. The oxidation state of the platinum ion must be 2+ to give a neutral salt.

19. (B) The complex ion is stable and remains together. There are three chlorides in this salt.

20. (E) There are no lone electron pairs in CH₄.

Set 3

Stoichiometry

1. Which sample of gas contains the same number of molecules as 68.0 g of hydrogen sulfide (H₂S)?
   - (A) 4.0 gram of hydrogen gas (H₂).
   - (B) 25.5 grams of ammonia gas (NH₃).
   - (C) 32.0 gram of oxygen gas (O₂).
   - (D) 66.0 grams of carbon dioxide gas (CO₂).
   - (E) 352 grams of uranium hexafluoride gas (UF₆).

2. Which 1.0 gram sample contains the greatest number of atoms?
   - (A) Aluminum chloride (AlCl₃)
   - (B) Ammonia (NH₃)
   - (C) Fluorine (F₂)
   - (D) Neon (Ne)
   - (E) Propane (C₃H₈)

3. Element M reacts with oxygen to produce a pure sample of MO₂. Find the atomic weight and the identity of M if 9.6 g of oxygen reacts with 16.5 g of M to produce 26.1 g of MO₂.
   - (A) 12 (C)
   - (B) 14 (N)
   - (C) 32 (S)
   - (D) 28 (Si)
   - (E) 55 (Mn)

4. Elements X and fluorine react to form two different compounds. In the first reaction 0.480 g of X reacts with 1.000 g of F. In the second 0.960 g of X reacts with 1.000 g of F. Which are possible formulas for the two compounds?

	First Compound	Second Compound
(A)	XF	X₂F₂
(B)	XF₂	XF
(C)	XF₂	X₂F
(D)	X₂F	XF₂
(E)	X₂F₃	X₃F

5. Analysis of an organic compound revealed it contains only 0.566 g hydrogen, 2.641 g nitrogen, and 6.793 g carbon. What is the empirical formula of the compound?
   - (A) C₃H₃N
   - (B) C₆H₆N
   - (C) C₁₂HN₅
   - (D) CH₃CH₂NH₂
   - (E) HCN

6. A compound has the empirical formula C₃H₄O and a molecular weight of 168.2. Its molecular formula is:
   - (A) C₃H₄O
   - (B) C₉H₁₄NO₂
   - (C) C₁₂H₁₆O
   - (D) C₉H₁₂O₃
   - (E) C₈H₈O₄

7. When 1.24 g of an organic compound with the formula $C_xH_yO_z$ is burned in excess oxygen, 1.76 g of carbon dioxide (CO_2) and 1.08 g of water vapor (H_2O) are obtained. What is the empirical formula of the compound?
   - (A) CHO
   - (B) CH_2O
   - (C) CH_3O
   - (D) $C_2H_3O_4$
   - (E) C_3H_5O

8. When germanium (Ge) is reacted with an excess of oxygen, germanium oxide is formed. If the reaction vessel plus the GeO_2 formed in this reaction weighs 43.78 g after the reaction, and it has gained 8.90 g during the reaction, what is the weight of the empty reaction vessel?
   - (A) 8.90 g
   - (B) 14.69 g
   - (C) 23.60 g
   - (D) 30.96 g
   - (E) 34.88 g

9. Octane, (C_8H_{18}, MW=114.08) is burned completely in an excess of air to give carbon dioxide, (CO_2), and water vapor (H_2O). How many grams of carbon dioxide can be produced by the complete combustion of 28.52 g of octane?
   - (A) 11.0 g
   - (B) 22.0 g
   - (C) 44.0 g
   - (D) 88.0 g
   - (E) 176.0 g

10. When 0.500 mol of $AlCl_3$(aq) is mixed with 0.300 mol of K_2CrO_4(aq), what is the maximum number of moles of $Al_2(CrO_4)_3$(s) that can be formed?
    - (A) 0.100 mol
    - (B) 0.250 mol
    - (C) 0.500 mol
    - (D) 0.600 mol
    - (E) 0.900 mol

11. How many milliliters of 6.0 M hydrochloric acid are required to prepare 250. mL of a 1.20 M HCl(aq) solution?
    - (A) 200. mL
    - (B) 125. mL
    - (C) 50.0 mL
    - (D) 28.8 mL
    - (E) 12.5 mL

12. The density of a solution of 20.0 % (by weight) nitric acid, HNO_3 (FW = 63.0), solution in water is 1.117 g/mL (1.117 kg/L).
    What is the molarity of this solution?
    - (A) 88.8 M
    - (B) 3.54 M
    - (C) 3.17 M
    - (D) 2.84 M
    - (E) 1.13 M

13. A 12.0 % solution of KOH(aq) (FW = 56.11) has a density of 1.1079 g/mL. Which set of answers is the mole fraction, molality and molarity respectively for this solution?
    - (A) 0.042, 2.37, 2.43
    - (B) 0.042, 2.43, 2.37
    - (C) 2.43, 0.042, 2.37
    - (D) 2.43, 2.37, 0.042
    - (E) 2.37, 2.43, 0.042

14. To determine the molar concentration of a sulfuric acid (H_2SO_4) solution, aqueous $Ba(OH)_2$ is added in excess, assuring complete reaction, to 20.0 mL of the acid. The resulting pure precipitate, $BaSO_4$ (FW=233.4), weighs 0.560 g. What was the concentration of the acid?
    (A) 6.54 M
    (B) 0.480 M
    (C) 0.240 M
    (D) 0.120 M
    (E) 0.060 M

15. When 30.0 mL of 0.100 M iron (III) nitrate, $(Fe(NO_3)_3)$ is reacted with a 0.200 M potassium oxalate, $(K_2C_2O_4)$ solution, iron (III) oxalate is precipitated. What is the minimum volume of $K_2C_2O_4$(aq) required for the maximum yield of $Fe_2(C_2O_4)_3$ precipitate?
    (A) 15.0 mL
    (B) 20.0 mL
    (C) 22.5 mL
    (D) 40.0 mL
    (E) 60.0 mL

16. What is the molarity of chloride ion, (Cl^-(aq)), in 100. mL of a 0.500 M barium chloride, ($BaCl_2$) solution?
    (A) 0.050 M
    (B) 0.100 M
    (C) 0.250 M
    (D) 0.500 M
    (E) 1.00 M

17. The molecular weight is determined by dissolving 5.00 g of a compound in 50.0 g of benzene. The freezing point was lowered by 2.5°C. The freezing point constant (K_f) for benzene is 5.0 °C molal^{-1}. What is the molecular weight?
    (A) 25
    (B) 50.
    (C) 100.
    (D) 150.
    (E) 200.

18. Methanol (CH_3OH) is more effective than an equal weight of ethanol (CH_3CH_2OH) in lowering the freezing point of 100. g of water because the methanol
    (A) is more soluble in water.
    (B) has a lower boiling point.
    (C) forms stronger hydrogen bonds.
    (D) has more dissolved particles.
    (E) has a higher vapor pressure.

19. A solution of a nonvolatile, nonionizing solute in toluene is prepared. Which is true?

	Vapor pressure	Melting point	Boiling point
(A)	Increased	Increased	Decreased
(B)	Decreased	Decreased	Increased
(C)	Increased	Increased	Increased
(D)	Decreased	Increased	Increased
(E)	Decreased	Decreased	Decreased

20. What will be the vapor pressure at 25°C of a 30.0% (by weight) solution of glucose, $C_6H_{12}O_6$ (MW=180), in water? The vapor pressure of water at 25°C is 23.8 mm Hg.
    (A) 7.1 mm Hg
    (B) 10.2 mm Hg
    (C) 16.7 mm Hg
    (D) 22.8 mm Hg
    (E) 23.3 mm Hg

Answer Key—Set 3

1.	(A)	6.	(D)	11.	(C)	16.	(E)
2.	(E)	7.	(C)	12.	(B)	17.	(E)
3.	(E)	8.	(B)	13.	(B)	18.	(D)
4.	(B)	9.	(D)	14.	(D)	19.	(B)
5.	(A)	10.	(A)	15.	(C)	20.	(D)

Explanations

1. **(A)** There are two moles of H_2S (FW=34.0) in 68.0 grams. The H_2 (4.0 g ÷ 2.0 g mole^{-1}) also has 2.0 moles of molecules.

2. **(E)** The number of moles is 1.0 g divided by the molar mass and multiplied by the number of atoms in the formula. Propane has $((1.0 \div 44.0) \times 11) = 0.25$ moles of atoms.

3. **(E)**
n_{oxygen} = 0.60 mol
n_M = $\frac{1 \text{ mol M}}{2 \text{ mol O}} \times 0.60 \text{ mol O}$ = 0.30 mol
The atomic weight of M = $\frac{16.5 \text{ g}}{0.30 \text{ mole}}$ = 55 g mol^{-1}

4. **(B)** The weights suggest that the second compound has twice as many atoms of X per atom of fluorine as the first. XF has half as many atoms of F per atom of X as XF_2. This is another way to say the same thing.

5. **(A)**

Element	Weight	Moles	Ratio
Carbon	6.793 g	0.566	3.00
Hydrogen	0.566 g	0.566	3.00
Nitrogen	2.641 g	0.189	1.00

6. **(D)** C_3H_4O has a formula weight of 56.03. The multiple is 3 (168.2 ÷ 56.03 = 3.005). $(C_3H_4O)_3 = C_9H_{12}O_3$

7. **(C)**

Element	Mol	Weight	Mol	Ratio
Carbon	0.040	0.48 g	0.040	1.0
Hydrogen	0.120	0.12 g	0.120	3.0
Oxygen	NA	0.64 g	0.040	1.0
Total		1.24 g		

Stoichiometry 107

8. **(B)**

	Ge	+ O$_2$	⇒ GeO$_2$
Weight	20.18 g	8.90 g	29.09 g
Mol	0.278	0.278	0.278

Wt$_{vessel}$ = 43.78 g - 29.09 g = 14.69 g

9. **(D)** C$_8$H$_{18}$ + 12½ O$_2$ ⇒ 8CO$_2$ + 9H$_2$O

    n$_{octane}$ = 28.52 g ÷ 114.08 g mol^{-1} = 0.2500

    n$_{carbon\ dioxide}$ = $\dfrac{8\ mol\ CO_2}{1\ mol\ C_8H_{18}}$ × n$_{octane}$

    n$_{carbon\ dioxide}$ = 8 × 0.2500 = 2.000 mol

    wt$_{carbon\ dioxide}$ = 2.000 × 44.0 = 88.0 g

10. **(A)** 2Al^{3+} + 3CrO$_4^{2-}$ ⇒ Al$_2$(CrO$_4$)$_3$

    n$_{aluminum\ chromate}$ = ½ n$_{aluminum\ ion}$
    n$_{aluminum\ chromate}$ = ½ × 0.500 = 0.250 mol
    n$_{aluminum\ chromate}$ = ⅓ n$_{chromate\ ion}$
    n$_{aluminum\ chromate}$ = ⅓ × 0.300 = 0.100 mol

    The chromate ion is the limiting reactant, and 0.100 mol is the maximum that can be formed.

11. **(C)** The dilution equation is: M$_1$V$_1$ = M$_2$V$_2$

    n$_{HCl}$ = 0.300 mol

    V$_{HCl}$ = $\dfrac{0.250\ L \times 1.20\ mol\ L^{-1}}{6.0\ mol\ L^{-1}}$ = 0.050 L = 50. mL

12. **(B)** Basis: 1000. g$_{solution}$ = 1.000 kg$_{solution}$

    n$_{nitric\ acid}$ = 200. g ÷ 63.0 g mol^{-1} = 3.17 mol
    L$_{solution}$ = 1.000 kg$_{solution}$ ÷ 1.117 kg L^{-1} = 0.895 L
    Molarity = 3.17 mol ÷ 0.895 L = 3.54 M

13. **(B)** Basis: 1000. g$_{solution}$ = 1.000 kg$_{solution}$

    n$_{KOH}$ = 120. g ÷ 56.0 g mol^{-1} = 2.14 mol
    n$_{HOH}$ = 880. g ÷ 18.0 g mol^{-1} = 48.8 mol
    x = 2.14 ÷ (2.14 + 48.8) = 0.042
    Molality = $\dfrac{2.14\ mol\ KOH}{0.880\ kg\ HOH}$ = 2.43 m

    L$_{solution}$ = $\dfrac{1000\ kg_{solution}}{1.1079\ kg\ L^{-1}}$ = 0.903 L

    Molarity = 2.14 mol ÷ 0.903 L = 2.37 M

Note that:
    mole fraction (x) << molarity (M) or molality (m)
    molarity (M) < molality (m)

14. **(D)** $Ba(OH)_2(aq) + H_2SO_4(aq) \Rightarrow BaSO_4(s) + 2H_2O(l)$

    $n_{barium\ sulfate}$ = 0.560 g ÷ 233.4 g mol^{-1}
    $n_{barium\ sulfate}$ = 0.00240 mol
    $n_{sulfuric\ acid}$ = $n_{barium\ sulfate}$
    Molarity = 0.00240 mol ÷ 0.0200 L = 0.120 $\underline{M}$

15. **(C)** $2Fe(NO_3)_3(aq) + 3K_2C_2O_4(aq) \Rightarrow Fe_2(C_2O_4)_3(s) + 6KNO_3(aq)$

    $n_{iron\ nitrate}$ = 0.0300 L × 0.100 mol L^{-1}
    $n_{iron\ nitrate}$ = 0.00300 mol
    $n_{potassium\ oxalate}$ = $\dfrac{3\ mol\ K_2C_2O_4}{2\ mol\ Fe(NO_3)_3}$ × $n_{iron\ nitrate}$
    = ½ × 0.00300 = 0.00450 mol
    $V_{potassium\ oxalate}$ = $\dfrac{0.00450\ mol}{0.20\ mol\ L^{-1}}$ = 0.0225 L
    $V_{potassium\ oxalate}$ = 22.5 mL

16. **(E)** $BaCl_2(aq) \Rightarrow Ba^{2+}(aq) + 2Cl^{1-}(aq)$

    $n_{chloride}$ = 2 × $n_{barium\ chloride}$
    The same volume will contain twice the moles of chloride ion.
    Molarity = 2 × 0.500 $\underline{M}$ = 1.00 $\underline{M}$

17. **(E)**

    Weight = 5.00 g$_{solute}$ ÷ 0.050 kg$_{benzene}$ = 100. g kg^{-1}
    molality = $\Delta T \div K_f$ = 2.5°C ÷ 5.0°C m^{-1} = 0.50 mol kg^{-1}
    Molar mass = $\dfrac{100.\ g\text{-}kg^{-1}}{0.50\ mol\ kg^{-1}}$ = 200 g mol^{-1}

18. **(D)** The same weight of methanol will contain more moles of molecules because of its lower molecular weight.

19. **(B)** Solutions have lower vapor pressures than the corresponding solvents due to increased intermolecular attractions. This lower vapor pressure lowers the freezing (melting) point and raises the boiling point.

20. **(D)** Basis: 100. g of solution

    n_{solute} = $\dfrac{30.0\ g}{180\ g\ mol^{-1}}$ = 0.167 mol
    $n_{solvent}$ = $\dfrac{70.0\ g}{18.0\ g\ mol^{-1}}$ = 3.89 mol
    Mol fraction$_{solvent}$ = $\dfrac{3.89\ mol}{(3.89\ mol + 0.17\ mol)}$ = 0.958
    $P_{solvent}$ = $x_{solvent} P°_{solvent}$ = 0.958 × 23.8 mm Hg
    $P_{solvent}$ = 22.8 mm Hg

Set 4

States of Matter

1. Which is a possible explanation why silicon carbide, SiC, is twice as hard as zinc oxide, ZnO, even though they both have a hexagonal crystal lattice? Silicon carbide:
   (A) has atoms which fit into the lattice more neatly.
   (B) has higher van der Waals forces.
   (C) has higher electrostatic forces.
   (D) is a covalent network solid.
   (E) is a stronger dipole.

2. The spontaneous endothermic solution of a salt in water suggests that:
   (A) all salts release energy when they dissolve.
   (B) all salts absorb energy when they dissolve.
   (C) the bond energy is more than the hydration energy.
   (D) the bond energy is less than the hydration energy.
   (E) the bond energy is equal to the hydration energy.

3. Molecular liquids:
   (A) boil at low temperatures
   (B) generally have high vapor pressures
   (C) dissolve readily in polar solvents
   (D) have covalent bonds between the molecules
   (E) conduct an electric current

4. Which explains the trend in the melting temperatures of the four molecules?
   (A) CI_4 is polar.
   (B) CF_4 is ionic.
   (C) CF_4 is hydrogen bonded.
   (D) CI_4 is a network solid.
   (E) CI_4 has more electrons.

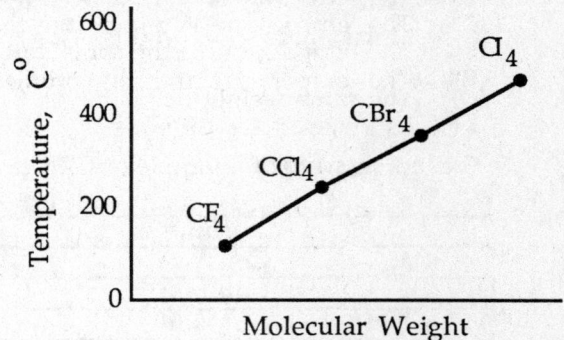

5. During the heating of a solid <u>at its melting temperature</u> the:
   (A) average kinetic energy is constant, but the average potential energy increases.
   (B) average potential energy is constant, but the average kinetic energy increases.
   (C) average potential and kinetic energies remain the same.
   (D) average kinetic energy of the liquid phase is greater than the solid phase.
   (E) average potential energy of the solid phase is greater than the liquid phase.

Questions 6 and 7 both refer to the diagram at the right.

6. What is the point that represents the temperature above which the gas will not condense at a high pressure?
   - (A) A
   - (B) B
   - (C) C
   - (D) D
   - (E) E

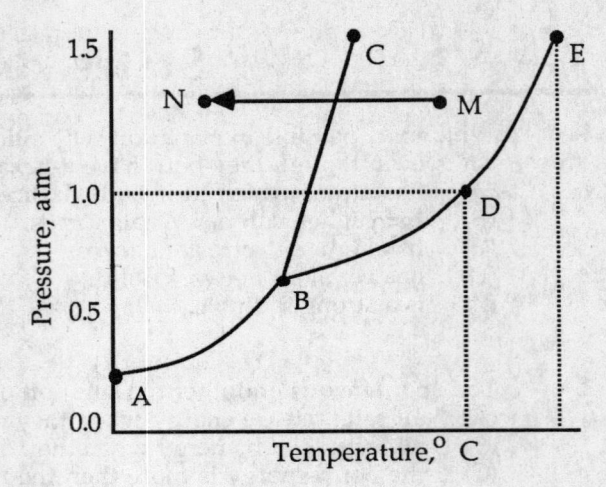

7. A phase change that occurs from point M to point N is:
   - (A) condensation.
   - (B) crystallization.
   - (C) boiling.
   - (D) sublimation.
   - (E) vaporization.

8. The conditions of a confined gas are changed. It is observed that the pressure is higher, but the molecules hit the walls of the container with the original force. Which is true?
   - (A) The temperature is the same, but the volume is larger.
   - (B) The temperature and the volume are the same.
   - (C) The temperature is higher, but the volume is the same.
   - (D) The temperature is the same, but the volume is smaller.
   - (E) The temperature and the volume are both smaller.

9. The temperature of a sample of CO_2 gas is increased. Which can be true?

	Volume	Pressure	Density
(A)	Constant	Increased	Increased
(B)	Constant	Increased	Constant
(C)	Constant	Constant	Decreased
(D)	Increased	Increased	Constant
(E)	Increased	Increased	Increased

10. A correction must be made for intermolecular attractions at low temperatures and/or high pressures. If 'a' symbolizes this factor, which is the gas law equation that corrects for 'a' for one mole of a gas.
    - (A) $(P + (a/V^2))V = RT$
    - (B) $(P - (a/V^2))V = RT$
    - (C) $P(V - a) = RT$
    - (D) $P(V + a) = RT$
    - (E) $PV = RT$

11. Two flasks containing the same inert gas are at the same temperature and pressure of 800. mm Hg. One flask has a volume of 1.0 L and the other a volume of 2.0 L. Enough volatile liquid is injected into each of the flasks to allow phase equilibrium to be established. No leakage occurs. If the pressure in the 1.0 L flask is 900 mm Hg, what is the pressure in the 2.0 L flask?
    (A) 1000. mm Hg.
    (B) 900. mm Hg.
    (C) 850. mm Hg.
    (D) 800. mm Hg.
    (E) 450. mm Hg.

12. A solid at 100.°C was sealed in a vessel containing dry air. After one hour the temperature had decreased to 80°C and the pressure had increased. It can be concluded from these data that the solid may have:
    (A) adsorbed some component of the air.
    (B) reacted with the oxygen in the air.
    (C) sublimed.
    (D) boiled.
    (E) melted.

13. A gas has a density of 2.68 g–L^{-1} at a pressure of 1.50 atm and a temperature of 27°C. What is its molecular weight?
    (A) 44 g mol^{-1}
    (B) 59 g mol^{-1}
    (C) 66 g mol^{-1}
    (D) 89 g mol^{-1}
    (E) 109 g mol^{-1}

14. One gram (1.00 g) of a gaseous hydrocarbon occupies 0.821 L at 1.00 atm. and 147°C. The compound is:
    (A) CH_4
    (B) C_2H
    (C) C_2H_4
    (D) C_3H_6
    (E) C_4H_{82}

15. A 1.00 L container at 273°C contains 6.00 moles of helium gas. What is the pressure of the gas?
    (A) 7.5 atm.
    (B) 22.4 atm.
    (C) 89.6 atm.
    (D) 134 atm.
    (E) 269 atm.

16. A mixture of nitrogen, N_2, and oxygen, O_2, occupies a 1.0 L volume at a pressure of 800. mm Hg. There are 4 times as many nitrogen molecules as oxygen molecules.
    All the N_2 molecules are removed and placed in a different 1.0 L container at the same temperature. What would be the pressure in the new container, assuming ideal behavior?
    (A) 800. mm Hg
    (B) 640. mm Hg
    (C) 600. mm Hg
    (D) 200. mm Hg
    (E) 160. mm Hg

17. Which is necessary to determine the vapor pressure of water in a gas which has been collected by the displacement of water in a gas measuring tube (eudiometer)?
    (A) The volume of the gas.
    (B) The barometric pressure.
    (C) The volume of the water.
    (D) The water solubility of the gas.
    (E) The temperature of the water.

18. The vapor pressure of hexane, C_6H_{14}, is 100. mm Hg at 16°C. What is the vapor pressure at 16°C of a solution containing 0.500 mol of nonvolatile solute dissolved in 4.50 mol of heptane?
    (A) 180. mm Hg
    (B) 110. mm Hg
    (C) 100. mm Hg
    (D) 90.0 mm Hg
    (E) 10.0 mm Hg

19. A Beaker 'A' contains distilled water and Beaker 'B' contains an equal volume of a 1.0 m solution of sugar in water. They are placed side–by–side in a sealed small aquarium. Which will be true one day later?
    (A) Beaker 'B' will contain more liquid than Beaker 'A'.
    (B) Beaker 'A' will contain more liquid than Beaker 'B'.
    (C) The sugar will be less than 1.0 m in Beaker 'B'.
    (D) The sugar will be more than 0.0 m in Beaker 'A'.
    (E) Both beakers contents will be the same as they started.

20. Methane, CH_4, effused through an small hole at a rate of 5.00 mL sec^{-1}. What would be the rate, at the same temperature and through the same hole, that sulfur trioxide, SO_3, would effuse?
    (A) 25.0 mL sec^{-1}
    (B) 11.2 mL sec^{-1}
    (C) 5.00 mL sec^{-1}
    (D) 2.23 mL sec^{-1}
    (E) 1.00 mL sec^{-1}

States of Matter 113

Answer Key—Set 4

1.	(D)	6.	(E)	11.	(B)	16.	(B)
2.	(C)	7.	(B)	12.	(C)	17.	(E)
3.	(A)	8.	(D)	13.	(A)	18.	(D)
4.	(E)	9.	(B)	14.	(D)	19.	(A)
5.	(A)	10.	(A)	15.	(E)	20.	(B)

Explanations

1. **(D)** The elements of Group IVA sometimes form three-dimensional network covalancies. Silicon carbide, SiC, is used in the manufacture of grinding wheels and sandpaper.

2. **(C)** When a salt dissolves in water, energy is absorbed as the crystal lattice is broken, and energy is released as the ions hydrate to the water molecules. The energy of the bond breaking exceeds the energy of the bond formation for an endothermic reaction (net energy absorbed).

3. **(A)** Molecular liquids have low boiling points. The molecules in molecular liquids are hydrogen bonded, are dipoles or are Van der Waals' attracted nondipoles. These are the weak attractions when compared to metallic, ionic and network bonding.

4. **(E)** The four molecules are all nondipoles held together by London dispersion forces. The forces are largest for the CI_4 molecule which have the highest number of electrons of the group.

5. **(A)** The temperature is constant at the melting temperature and therefore the average kinetic energy will not change. The average potential energy will increase due to the increased average distance between molecules in the liquid phase.

6. **(E)** Only the gaseous phase can exist above the critical temperature.

7. **(B)** The line BC represents the melting temperature curve.

8. **(D)** The molecules hitting the wall with the same force indicates velocity, the temperature and the kinetic energy are the same. Pressure and volume are inversely proportional. The volume must have been decreased to cause the increase in pressure.

9. **(B)** The increase in temperature will cause an increase in the pressure if the volume is constant. The density will be constant if the weight and volume are the same.

10. **(A)** The attraction between molecules will cause a decrease in the pressure. The correction factor must be added to the pressure.

11. **(B)** The vapor pressure of a liquid is dependent only on temperature.

12. **(C)** The sublimation of a solid to the gaseous phase both lowers the temperature (bond breaking is an endothermic reaction) and increases the pressure.

13. **(A)**
$$MW = \frac{DRT}{P} = \frac{2.68 \text{ g L}^{-1} \times 0.0821 \text{ L atm mol}^{-1} \text{ K} \times 300 \text{ K}}{1.50 \text{ atm}}$$
$$MW = 44.0 \text{ g mol}^{-1}$$

14. **(D)**
$$MW = \frac{gRT}{PV} = \frac{1.00 \text{ g L}^{-1} \times 0.0821 \text{ L atm mol}^{-1} \text{ K} \times 420 \text{ K}}{1.00 \text{ atm} \times 0.821 \text{ L}}$$
$$MW = 42.0 \text{ g mol}^{-1} \quad \text{which matches } C_3H_6 \text{ (FW = 42.0)}$$

15. **(E)**
$$P = \frac{nRT}{V} = \frac{6.00 \text{ mol} \times 0.0821 \text{ L atm mol}^{-1} \text{ K} \times 546 \text{ K}}{1.00 \text{ L}}$$
$$P = 269.0 \text{ atm.}$$

16. **(B)** The partial pressure of the nitrogen is $4/5$ or 0.800 of the 800. mm Hg pressure.
$P_{nitrogen} = 0.800 \times 800. \text{ mm Hg} = 640. \text{ mm Hg}$

17. **(E)** The vapor pressure of a liquid is dependent only on temperature.

18. **(D)** Raoult's Law predicts the vapor pressure of the solution.

$P_{solvent} = x_{solvent} P°_{solvent}$
$x_{solvent} = \dfrac{4.50 \text{ mol}}{(0.50 + 4.50)} = 0.900 \text{ mol}$
$P_{solvent} = 0.900 \times 100. \text{ mm Hg} = 90.0 \text{ mm Hg}$

19. **(A)** Water solutions have lower vapor pressures than pure water because of the attractions between the solute and the solvent are larger than between water molecules. The sugar solution and the pure water will evaporate until vapor–liquid equilibrium is established at the temperature of the aquarium. But, more of the pure water will evaporate because of the higher vapor pressure.

20. **(B)** Graham's law predicts the effusion rate.
$$\frac{\text{Rate of SO}_3}{\text{Rate of CH}_4} = \frac{(MW\ CH_4)^{1/2}}{(MW\ SO_3)^{1/2}}$$

$$\text{Rate of SO}_3 = \frac{(16)^{1/2}}{(80)^{1/2}} \times 5.00 \text{ mL sec}^{-1}$$
$$\text{Rate of SO}_3 = 2.23 \text{ mL sec}^{-1}$$

Set 5

Reaction Kinetics

1. The reaction:

    $$2NO(g) + Br_2(g) \Rightarrow 2NOBr(g)$$

    is second order with respect to NO and first order with respect to Br_2. Which is the rate law equation?

    (A) $+\dfrac{\Delta[NO]}{\Delta t} = +\dfrac{\Delta[Br_2]}{\Delta t} = k[NO]^2[Br_2]$

    (B) $-\dfrac{1}{2}\dfrac{\Delta[NOBr]}{\Delta t} = +\dfrac{\Delta[Br_2]}{\Delta t} = k[NO]^2[Br_2]$

    (C) $+\dfrac{1}{2}\dfrac{\Delta[NOBr]}{\Delta t} = -\dfrac{1}{2}\dfrac{\Delta[NO]}{\Delta t} = k[NOBr]^2$

    (D) $-\dfrac{\Delta[NO]}{\Delta t} = -\dfrac{1}{2}\dfrac{\Delta[Br_2]}{\Delta t} = k[NOBr]^2$

    (E) $-\dfrac{1}{2}\dfrac{\Delta[NO]}{\Delta t} = -\dfrac{\Delta[Br_2]}{\Delta t} = k[NO]^2[Br_2]$

2. For the reaction: $2O_3(g) \Rightarrow 3O_2(g)$

    the reaction rate of O_3 was found to be -2.0 mm Hg sec^{-1}. What was the rate for the O_2 during the same time period?
    (A) 3.0 mm Hg sec^{-1}
    (B) 2.0 mm Hg sec^{-1}
    (C) 1.5 mm Hg sec^{-1}
    (D) 1.3 mm Hg sec^{-1}
    (E) 0.67 mm Hg sec^{-1}

Questions 3-7 refer to these data for the reaction:

$$2NO(g) + 2H_2(g) \Rightarrow N_2(g) + 2H_2O(g)$$

Run	Pressure(NO)	Pressure(H_2)	Rate
1	0.375 atm	0.500 atm	6.43×10^{-4} atm sec^{-1}
2	0.375	0.250	3.15×10^{-4}
3	0.188	0.500	1.56×10^{-4}
4	1.000	1.000	9.00×10^{-3}

3. What is the rate law equation for this reaction?
    (A) Rate $= k[P(NO)]$
    (B) Rate $= k[P(NO)]^2[P(H_2)]$
    (C) Rate $= k[P(NO)][P(H_2)]^2$
    (D) Rate $= k[P(NO)][P(H_2)]$
    (E) Rate $= k[P(NO)]^2$

4. If both the NO and the H$_2$ were initially at 2.000 atm, the initial rate of the reaction would be:
   (A) 9.00 x 10^{-3} atm sec^{-1}
   (B) 1.80 x 10^{-2} atm sec^{-1}
   (C) 2.70 x 10^{-2} atm sec^{-1}
   (D) 3.60 x 10^{-2} atm sec^{-1}
   (E) 7.20 x 10^{-2} atm sec^{-1}

5. The value of the rate constant, k, is:
   (A) 9.00 x 10^{-3} atm sec^{-1}
   (B) 9.00 x 10^{-3} atm^{-1} sec^{-1}
   (C) 9.00 x 10^{-3} atm^{-2} sec^{-1}
   (D) 9.00 x 10^{-4} atm^3 sec
   (E) 9.00 x 10^{-4} atm^4 sec^{-1}

6. Which could be the rate determining step in the mechanism of this reaction?
   (A) 2NO + 2H$_2$ $\Rightarrow$ Intermediate
   (B) 2NO $\Rightarrow$ Intermediate
   (C) NO + 2H$_2$ $\Rightarrow$ Intermediate
   (D) NO + H$_2$ $\Rightarrow$ Intermediate
   (E) 2NO + H$_2$ $\Rightarrow$ Intermediate

7. What conclusion can be made from the data given?
   The rate determining step:
   (A) of a catalyzed reaction is faster.
   (B) is not assumed from the stoichiometry of the reaction.
   (C) is the fastest step.
   (D) is the slowest step.
   (E) at a temperature 10°C higher doubles the rate.

Questions 8. and 9. are for the reaction:

$$N_2(g) + O_2(g) \Rightarrow 2NO(g)$$

Assume there are no intermediate steps or complications. Use the choices for both questions 8 and 9.

(A) The reaction rate will increase, but not double.
(B) The reaction rate will double.
(C) The reaction rate will more than double.
(D) The reaction rate will decrease, but not halve.
(E) The reaction rate will halve.

8. The temperature is adjusted so the reaction just begins. If the pressure of the O$_2$ is doubled at this temperature: .....

9. The number of molecules and the volume are held constant. If the temperature is raised high enough to double the number of collisions: .....

10. Given these data for the reaction: I⁻ + OCl⁻ ⇒ IO⁻ + Cl⁻

Run	[I⁻] M	[OCl⁻] M	Rate
1	0.07 M	0.21 M	0.0537 mol L⁻¹ sec⁻¹
2	0.14	0.42	0.215
3	0.28	0.42	0.429
4	0.56	0.21	0.429

What is the rate equation for this reaction?
(A) Rate = k [I⁻]
(B) Rate = k [OCl⁻]
(C) Rate = k [I⁻] [OCl⁻]
(D) Rate = k [I⁻]²[OCl⁻]
(E) Rate = k [I⁻] [OCl⁻]²

11. The mechanism for a reaction is:
Step 1:   A + B   ⇔ C
Step 2:   C + B   ⇒ D
Step 3:   D + A   ⇒ F
Overall: 2A + 2B ⇒ F

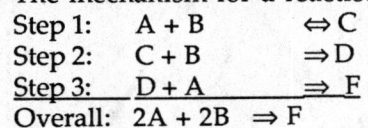

The energy diagram for the same reaction is:

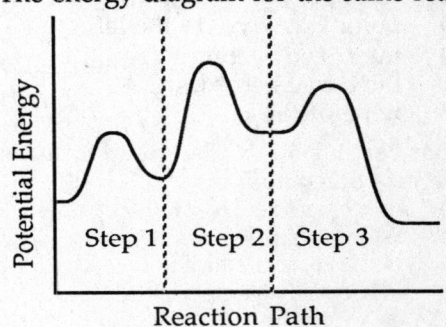

Which is true about this system and these data?
(A) Step 3 is the fastest step.
(B) A catalyst would affect the potential energy of each step equally.
(C) The rate law equation is: Rate = k [A]²[B]²
(D) The rate law equation is: Rate = k [A] [B]²
(E) The activation energy of each step is equal.

12. For the reaction: $H_2(g) + Cl_2(g) \Rightarrow 2HCl(g)$

   a. Equal volumes of hydrogen and chlorine react explosively.
   b. A large excess of either will not explode, but may react.

What would explain this difference?
(A) An increased activation energy.
(B) A decreased heat of reaction.
(C) An increased concentration of the chain–maintaining agent.
(D) An increased concentration of the chain–ending reactant.
(E) A decreased collision probability between chain–maintaining species.

13. The decomposition of NaClO$_3$ is a first order reaction. In an experiment, a sample of NaClO$_3$ was 90% decomposed in 48.0 minutes. How long would it take for a sample to be 50% decomposed?
   (A) 14.4 minutes
   (B) 21.6 minutes
   (C) 24.0 minutes
   (D) 43.2 minutes
   (E) 48.0 minutes

Questions 14. and 15. refer to the graph shown.
The solid line (B) is a graph of the kinetic energy distribution at a given temperature. Point E$_a$ represents the activation energy.

Use the set of choices for both questions.

   (A) point E$_a$ moves to the right.
   (B) point E$_a$ moves to the left.
   (C) the curve becomes A.
   (D) the curve becomes C.
   (E) none of these.

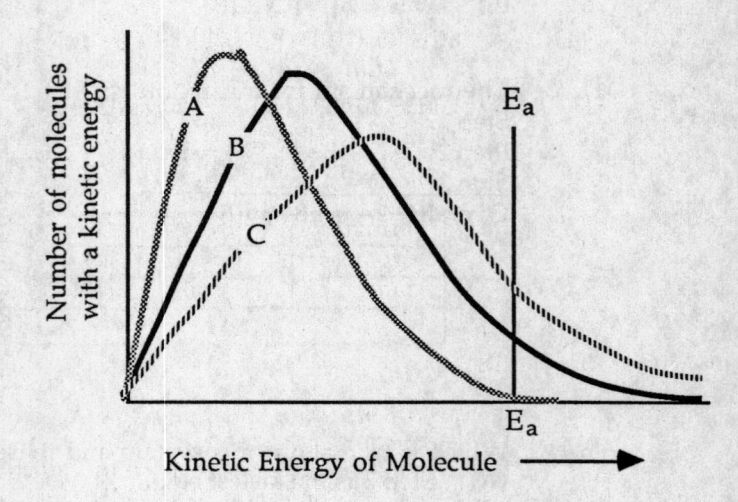

14. If the temperature is increased.....

15. If a catalyst is added.....

16. The energy diagram is for the reaction:

   $$NO(g) + O_3(g) \Leftrightarrow NO_2(g) + O_2(g)$$

   The distance marked 'X' represents:
   (A) The activation energy for:
       NO(g) + O$_3$(g) ⇔ NO$_2$(g) + O$_2$(g)
   (B) The heat of reaction for:
       NO(g) + O$_3$(g) ⇔ NO$_2$(g) + O$_2$(g)
   (C) The activation energy for:
       NO$_2$(g) + O$_2$(g) ⇔ NO(g) + O$_3$(g)
   (D) The heat of reaction for:
       NO$_2$(g) + O$_2$(g) ⇔ NO(g) + O$_3$(g)
   (E) None of these.

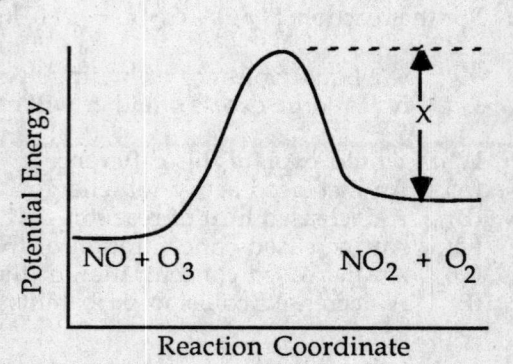

17. An approximate rule is that an increase in temperature of 10°C. will double the reaction rate. Which is doubled on the molecular level? The:
    (A) activation energy.
    (B) average kinetic energy.
    (C) average velocity of molecules.
    (D) fraction of molecules with the activation energy.
    (E) number of collisions.

18. What is the activation energy for a reaction where an increase in temperature from 27°C to 37°C exactly triples the initial rate of reaction?
    (A) 55.5 kcal
    (B) 41.0 kcal
    (C) 20.3 kcal
    (D) 18.1 kcal
    (E) 10.5 kcal

19. Given the mechanism of a hypothetical reaction:
    Step 1:   2A ⇔ intermediate$_{(1)}$ (fast equilibrium)
    Step 2:   intermediate$_{(1)}$ + B ⇒ intermediate$_{(2)}$ (slow)
    Step 3:   intermediate$_{(2)}$ + B ⇒ A$_2$B$_2$ (fast)
    Overall: 2A + 2B ⇒ A$_2$B$_2$

    Which is the rate law equation?
    (A) rate = k[A]^2
    (B) rate = k[B]^2
    (C) rate = k[A][B]
    (D) rate = k[A]^2[B]
    (E) rate = k[A][B]^2

20. The reaction 2NO(g) + Cl$_2$(g) ⇒ 2NOCl(g) has the rate law:

    Rate = k[NO]^2[Cl$_2$]

    Which is a possible mechanism?

(A)	2NO	⇒ N$_2$O$_2$	(slow)
	N$_2$O$_2$ + Cl$_2$	⇒ 2NOCl	(fast)
(B)	2NO	⇔ N$_2$O$_2$	(fast equilibrium)
	N$_2$O$_2$	⇒ NO$_2$ + N	(slow)
	NO$_2$ + N + Cl$_2$	⇒ 2NOCl	(fast)
(C)	Cl$_2$	⇒ 2Cl	(slow)
	2NO	⇔ N$_2$O$_2$	(fast equilibrium)
	N$_2$O$_2$ + 2Cl	⇒ 2NOCl	(fast)
(D)	NO + Cl$_2$	⇔ NOCl$_2$	(fast equilibrium)
	NO + NOCl$_2$	⇒ 2NOCl	(slow)
(E)	NO + Cl$_2$	⇒ NOCl + Cl	(slow)
	Cl + NO	⇒ NOCl	(fast)

Answer Key—Set 5

1.	(E)	6.	(E)	11.	(D)	16.	(C)
2.	(A)	7.	(B)	12.	(E)	17.	(D)
3.	(B)	8.	(B)	13.	(A)	18.	(C)
4.	(E)	9.	(C)	14.	(D)	19.	(D)
5.	(C)	10.	(C)	15.	(B)	20.	(D)

Explanations

1. **(E)** There are 2NO's and one Br_2, so the rate of disappearance (minus sign is required) of the bromine, Br_2, is $1/2$ the [NO]. The order of [NO] and [Br_2] are 2 and 1 respectively.

2. **(A)** The reaction rate is $-(1/2 \times -2.0)$ or 1.0 mm Hg sec^{-1}. The appearance of O_2 is (3×1.0 mm Hg sec^{-1} = 3.0 mm Hg sec^{-1}.)

3. **(B)** Runs 1 and 3 show when P(NO) alone is doubled, the reaction rate is quadrupled. The order of NO is 2. Runs 1 and 2 show that when P(H_2) alone is doubled, the reaction rate is doubled. The order of H_2 is 1.

4. **(E)** The overall order is 3. The rate will be 8.000 times, $(2)^3$, that of run 4 at the same temperature.
Rate = $8.000 \times 9.00 \times 10^{-3} = 7.20 \times 10^{-2}$ atm sec-1

5. **(C)** Run 4 shows the rate constant, k, is 9.00×10^{-3}.

    Rate = $9.00 \times 10^{-3} \dfrac{atm}{sec} = k \dfrac{1}{atm^2\ sec} (1.0)^2 (1.0)\ atm^3$

6. **(E)** The rate determining step is the slow step, and the reactants will have coefficients which are equal to their orders in this step.

7. **(B)** The data only shows that the stoichiometry and the orders are different. All the other answers (except (C)) are essentially true, but are not reflected in the data.

8. **(B)** Doubling the pressure will double the collisions and therefore double the reaction rate.

9. **(C)** Raising the temperature high enough to double the number of collisions will also increase the effectiveness of the collisions.

Reaction Kinetics   121

10. **(C)** Runs 2 and 3 show that when [I⁻] is doubled, the reaction rate is doubled. The order of I⁻ is 1. Comparing runs 1 and 4 show that [OCl⁻] is 8–times larger and the reaction rate is 8–times larger. The order of OCl⁻ is 1.

11. **(D)** The highest activation energy is for step 2. This is the rate determining step.
Rate = $k_2[C][B]$.
Step is a fast equilibrium. $[C] = K_1[A][B]$. Substitute for [C] in the rate equation.
Rate = $k[A][B]^2$, where $k = K_1 k_2$.

12. **(E)** Diluting the system puts H_2 or Cl_2 molecules in between intermediate species (free atoms) essential for an explosive chain reaction step.

13. **(A)**
ln 0.10 = –k × 48.0 min
k = 0.0480 min⁻¹
$t_{1/2}$ = ln 0.50 ÷ 0.0480 min⁻¹ = 14.4 minutes

14. **(D)** Temperature shifts the temperature curve to the right and lower.

15. **(B)** A catalyst lowers the activation energy, E_a.

16. **(C)** 'X' is the reverse reaction activation energy.

17. **(D)** Increasing the temperature shifts the kinetic energy curve to the right, which increases (and sometimes doubles) the number of molecules with the required activation energy.

18. **(C)** Use the Arrhenius equation.

$$\ln 3.0 = \frac{-E_a}{1.987 \text{ cal mol}^{-1} \text{ K}^{-1}} \left[\frac{1}{310} - \frac{1}{300} \right]$$

19. **(D)** Step 2 is rate determining. Rate = $k_2 [Int_1][B]$. Step 1 is a fast equilibrium. $[Int_1] = K_1 [A]^2$. Substitute $K_1 [A]^2$ for $[Int_1]$ in the rate equation.

20. **(D)** For the slow step: rate = $k_2 [NOCl_2][NO]$.
The fast equilibrium gives: $[NOCl_2] = K_1 [NO][Cl_2]$.
Combining these will agree with the rate equation.

Set 6

Equilibrium

1. For a reaction involving only gases at 25°C., the equilbrium constant can be expressed in terms of molarity, K_c, or partial pressures K_p. Which is true about the numerical value of K_p?
   - (A) K_c is generally greater than K_p.
   - (B) K_c is generally less than K_p.
   - (C) K_c is generally equal to K_p.
   - (D) K_c is equal to K_p if the total moles of reactants and total moles of products are equal.
   - (E) K_c is greater than K_p if the total moles of reactants are less than the total moles of products.

2. The equilibrium reaction is:

   $$H_2(g) + I_2(g) \Leftrightarrow 2HI(g)$$

   When 2.00 mole each of hydrogen, H_2, and iodine, I_2, are reacted in a 1.00 L vessel, 3.50 mole of HI is produced.
   What is the value of the equilibrium constant, K_c?
   - (A) 196
   - (B) 56
   - (C) 49
   - (D) 14
   - (E) 5.4

3. The equilibrium constant for the reaction:

   $$SO_2(g) + NO_2(g) \Leftrightarrow SO_3(g) + NO(g)$$

   has a numerical value of 3.00 at a given temperature. Equimolar amounts SO_2 and NO_2 are reacted at this temperature and a total pressure of 3.00 atm.
   What percent of the SO_2 is converted to product?
   - (A) 87.5%
   - (B) 74.0%
   - (C) 63.4%
   - (D) 50.0%
   - (E) 33.0%

4. For which of the reactions will an increase in pressure cause a decrease in product (temperature remaining constant)?
   - (A) $N_2(g) + 3H_2(g) \Leftrightarrow 2NH_3(g)$
   - (B) $3Fe(s) + 4H_2O(g) \Leftrightarrow Fe_3O_4(s) + 4H_2(g)$
   - (C) $PCl_3(g) + Cl_2(g) \Leftrightarrow PCl_5(g)$
   - (D) $HCl(g) + H_2O(l) \Leftrightarrow H_3O^+(aq) + Cl^-(aq)$
   - (E) $CaCO_3(s) \Leftrightarrow CaO(s) + CO_2(g)$

5. Consider the equilibrium reaction:

$$2SO_2(g) + O_2(g) \Leftrightarrow 2SO_3(g)$$

What will be the effect of doubling the concentration of SO_3?
The concentration of:
(A) SO_2 and O_2 increase equally.
(B) SO_2 increases more than that of O_2.
(C) O_2 increases more than that of SO_2.
(D) SO_2 decreases more than that of O_2.
(E) O_2 decreases more than that of SO_2.

6. Barium hydroxide is completely ionized in dilute solution.

$$Ba(OH)_2(s) \Rightarrow Ba^{2+}(aq) + 2OH^-(aq)$$

Which is true?
(A) $[Ba^{2+}] = [OH^-]^2$
(B) $[Ba^{2+}] = 2 \times [OH^-]$
(C) $[OH^-] = [Ba^{2+}]^2$
(D) $[OH^-] = 2 \times [Ba^{2+}]$
(E) $[Ba^{2+}] = [Ba(OH)_2]$

7. The solubility of copper II iodate, $Cu(IO_3)_2$ in water is 3.27×10^{-3} M.
What is the K_{sp}?
(A) 2.14×10^{-5}
(B) 4.28×10^{-5}
(C) 1.40×10^{-7}
(D) 3.50×10^{-8}
(E) 7.00×10^{-8}

8. A 10.0 mL solution is 0.10 M with respect to both calcium nitrate, $Ca(NO_3)_2$, and magnesium nitrate, $Mg(NO_3)_2$. What happens if 0.10 M NaF is added by drops until precipitation occurs?

$K_{sp} = 4.0 \times 10^{-11}$ for CaF_2 at 25°C.

$K_{sp} = 6.4 \times 10^{-9}$ for MgF_2 at 25°C.
(A) The precipitate is calcium fluoride only.
(B) The precipitate is magnesium fluoride only.
(C) The precipitate contains both CaF_2 and MgF_2.
(D) The precipitate contains both $Ca(NO_3)_2$ and MgF_2.
(E) The precipitate contains both CaF_2 and $Mg(NO_3)_2$.

9. Why is hydroxide ion a strong base in water solutions?
(A) OH^- is the only base.
(B) OH^- is the conjugate base of the hydronium ion, H_3O^+.
(C) All bases react with water to produce OH^- ions.
(D) All acids react completely with OH^- ions.
(E) Strong bases dissolve completely in water and produce OH^- ions.

10. Which is true about a 0.25 M KOH solution?
    (A) [KOH] = 0.25 M
    (B) [K$^+$] = [OH$^-$] = 0.25 M
    (C) [KOH] = [K$^+$] = 0.25 M
    (D) [H$^+$] = [OH$^-$] = 1.0 × 10^{-7} M
    (E) pH = 0.60

11. The ionization reaction for water is:
    $2H_2O(l) \Leftrightarrow H_3O^+(aq) + OH^-(aq)$
    Which is a true statement about this reaction?
    (A) If water is added the [H$_3$O$^+$] and [OH$^-$] increase.
    (B) If acid is added the [H$_3$O$^+$] and [OH$^-$] decrease.
    (C) If base is added the [H$_3$O$^+$] and [OH$^-$] increase.
    (D) $K_w = \dfrac{[H_3O^+][OH^-]}{[H_2O]^2}$
    (E) In pure water [H$_3$O$^+$] equals [OH$^-$].

12. To what total volume of aqueous solution must 20.0 mL of 0.10 M HCl be diluted so that the pH is 1.30?
    (A) 20. mL
    (B) 25. mL
    (C) 40. mL
    (D) 50. mL
    (E) 110 mL

13. The term "K_a for the ammonium ion, NH$_4^+$," refers to which equation?
    (A) NH$_4^+$(aq) + H$_2$O(l) $\Leftrightarrow$ NH$_3$(aq) + H$_3$O$^+$(aq)
    (B) NH$_4^+$(aq) + OH$^-$(aq) $\Leftrightarrow$ NH$_3$(aq) + H$_2$O(l)
    (C) NH$_3$(aq) + H$_2$O(l) $\Leftrightarrow$ NH$_4^+$(aq) + OH$^-$(aq)
    (D) NH$_3$(aq) + H$_3$O$^+$(aq) $\Leftrightarrow$ NH$_4^+$(aq) + OH$^-$(aq)
    (E) None of these.

14. The K_a of hydrocyanic acid, HCN, is 5.0 × 10^{-10}. The pH of 0.05 M HCN(aq) is:
    (A) between 3.5 and 4.5
    (B) between 4.5 and 5.0
    (C) between 5.0 and 5.5
    (D) between 5.5 and 6.0
    (E) between 9.0 and 9.5

15. Benzoic acid, C$_6$H$_5$COOH, is 1.0% ionized in a 0.010 M solution.
    $C_6H_5COOH(aq) \Leftrightarrow H^+(aq) + C_6H_5COO^-(aq)$
    What is the value of K_a?
    (A) 1.0 × 10^{-8}
    (B) 1.0 × 10^{-6}
    (C) 1.0 × 10^{-4}
    (D) 1.0 × 10^{-3}
    (E) 1.0 × 10^{-2}

16. Which set of solutes will form a buffer when dissolved in water to make a Liter of solution?
    (A) 0.2 mole of NaOH with 0.2 mole of HCl.
    (B) 0.2 mole of NaCl with 0.2 mole of HNO$_3$.
    (C) 0.4 mole of CH$_3$COOH with 0.4 mole of NaOH.
    (D) 0.4 mole of NH$_3$ with 0.2 mole of HCl.
    (E) 0.3 mole of KOH with 0.2 mole of HBr.

17. An experiment is performed where it is desired to keep the pH at 3.70. Which system would you use?
    (A) 0.0002 M HCl
    (B) HCNO and CNO$^-$ with a K$_a$ = 2.19 x 10^{-4}
    (C) CH$_3$COOH and CH$_3$COO$^-$ with a K$_a$ = 1.74 x 10^{-5}
    (D) HOCl and OCl$^-$ with a K$_a$ = 2.95 x 10^{-8}
    (E) C$_6$H$_5$OH and C$_6$H$_5$O$^-$ with a K$_a$ = 5.00 x 10^{-11}

18. The K$_a$ for chloroacetic acid, CH$_2$ClOOH, is 1.35 x 10^{-3}. What is the pH of a solution that is 0.15 M in CH$_2$ClOOH and 0.25 M in sodium chloroacetate, CH$_2$ClOONa?
    (A) 1.17   (D) 3.09
    (B) 2.64   (E) 4.54
    (C) 2.87

19. What is the pH of a solution obtained by mixing 10. mL of 1.0 M HCl, 75 mL of water and 15 mL of 1.0 M NaOH?
    (A) 1.3   (D) 12.3
    (B) 1.7   (E) 12.7
    (C) 7.0

20. What is the pH of a solution obtained by titrating 50.0 mL of 0.100 M HNO$_2$ with 0.100 M NaOH to the equivalence point? The K$_a$ of HNO$_2$ is 4.5 x 10^{-4}.
    (A) 1.67   (D) 7.00
    (B) 2.35   (E) 8.02
    (C) 3.35

Answer Key—Set 6

1.	(D)	6.	(D)	11.	(E)	16.	(D)
2.	(A)	7.	(C)	12.	(C)	17.	(B)
3.	(C)	8.	(A)	13.	(A)	18.	(D)
4.	(E)	9.	(E)	14.	(C)	19.	(E)
5.	(B)	10.	(B)	15.	(B)	20.	(E)

Explanations

1. **(D)** $K_c = K_p(1/RT)^{\Delta n}$ or $K_p = K_c(RT)^{\Delta n}$
When $\Delta n = 0$ the $(1/RT)$ or the (RT) term drops out of the equation, and $K_c = K_p$

2. **(A)**

	$H_2(g)$	+ $I_2(g)$	$\Leftrightarrow$	$2HI(g)$
Start	2.00 M	2.00 M		0.0 M
Δ	-1.75	-1.75		+3.50
Finish	0.25	0.25		3.50

$K_c = \dfrac{(3.50)^2}{(0.25)^2} = 196$

3. **(C)**

	$CO_2(g)$ +	$H_2(g)$	$\Leftrightarrow$	$CO(g)$ +	$H_2O(g)$
Start	1.50 atm	1.50 atm		0.0 atm	0.0 atm
Δ	-x	-x		+x	+x
Finish	1.50-x	1.50-x		x	x

$\dfrac{x^2}{(1.50-x)^2} = 3.00$

$x = 0.951$

$\% = \dfrac{0.951}{1.50} \times 100 = 63.4\%$

4. **(E)** When pressure is increased the side of the reaction with the least moles of gas is favored. Reaction E has 0 moles of gaseous reactant and 1 mole of products. (E) will produce reactant at the expense of product as it returns to equilibrium.

5. **(B)** Increasing the concentration of a product shifts the equilibrium to the left. The concentrations of SO_2 and O_2 increase by stoichiometric (equal) amounts.

6. **(D)** The stoichiometry of the balanced equation predicts that for every Ba^{2+} formed two OH^- are produced.

7. **(C)** $Cu(IO_3)_2(s) \Leftrightarrow Cu^{2+}(aq) + 2IO_3^-(aq)$
$K_{sp} = [Cu^{2+}][IO_3^-]^2 = [3.27 \times 10^{-3}][6.54 \times 10^{-3}]^2$
$K_{sp} = 1.40 \times 10^{-7}$

Equilibrium 127

8. **(A)** The least soluble fluoride, and only the least soluble fluoride, precipitates first.
$$MF_2(s) \Leftrightarrow M^{2+}(aq) + 2F^-(aq)$$
where M^{2+} is Ca^{2+} or Mg^{2+}
K_{sp} = $[M^{2+}][F^-]^2$ = $[0.10][F^-]^2$
$[F^-]$ = $(K_{sp} \div 0.10)^{1/2}$
The CaF_2 preciptates when $[F^-]$ = 2.0×10^{-5} M.
The MgF_2 preciptates when $[F^-]$ = 2.5×10^{-4} M.

9. **(E)** Main group I hydroxides are strong bases which dissociate completely in water solutions to produce alkali metal ions and hydroxide ions. Other bases react with water to form hydroxide ions.

10. **(B)**

	KOH(s) $\Leftrightarrow$	K^+(aq)	+	OH^-(aq)
Start	0.25 M	0.0 M		0.0 M
Δ	-0.25	+0.25		+0.25
Finish	0.0	0.25		0.25

The $[K^+(aq)]$ and $[OH^-(aq)]$ are equal at 0.25 M.

11. **(E)** $2H_2O(l) \Leftrightarrow H_3O^+(aq) + OH^-(aq)$
At equilibrium,
$[H_3O^+(aq)] = [OH^-(aq)] = 1.0 \times 10^{-7}$ M.

12. **(C)** At pH = 1.30, $[H^+] = 10^{-1.30} = 0.050$ M.
For dilution problems:
$M_2V_2 = M_1V_1$
0.050 M × V_2 = 0.10 M × 20.0 mL
V_2 = 40. mL

13. **(A)** Another way to write the K_a equation is:
$NH_4^+(aq) \Leftrightarrow H^+(aq) + NH_3(aq)$
H_3O^+ is a hydrated proton, $H^+(aq)$.

14. **(C)**

	HCN(aq) $\Leftrightarrow$	H^+(aq)	+	CN^-(aq)
Finish	0.05 M	x M		x M

K_a = $\dfrac{x^2}{0.05}$ = 5×10^{-6}
pH = $-\log[H^+]$ = 5.3

15. **(B)**

	$C_6H_5COOH \Leftrightarrow$	H^+	+	$C_6H_5COO^-$
Finish	0.0100 M	1.00×10^{-4} M		1.00×10^{-4} M

K_a = $\dfrac{(1.00 \times 10^{-4})^2}{1.00 \times 10^{-2}}$ = 1.0×10^{-6}

16. **(D)** $NH_4^+(aq) \Leftrightarrow H^+(aq) + NH_3(aq)$

When acid is added to NH_3, some $NH_4^+(aq)$ is formed. A buffer is roughly equivalent concentrations of a weak acid and its conjugate base.

17. **(B)** The pH is 3.70 and the $[H^+] = 10^{-3.70} = 2.0 \times 10^{-4}$.

$K_a = [H^+] \times \dfrac{[\cancel{HA}]}{[\cancel{A^-}]} = 2.0 \times 10^{-4}$.

The HCNO and CNO⁻ system comes closest to this K_a.

18. **(D)** $pH = pK_a + \log\dfrac{[A^-]}{[HA]} = 2.87 + \log\dfrac{(0.25)}{(0.15)} = 3.09$

19. **(E)** $[OH^-] = \dfrac{15 \text{ mL} \times 1.0 \text{ mol L}^{-1}}{100. \text{ mL}} = 0.015 \underline{M}$

$[H^+] = \dfrac{10 \text{ mL} \times 1.0 \text{ mol L}^{-1}}{100. \text{ mL}} = 0.010 \underline{M}$

	$H_2O(l)$	$\Leftrightarrow$ $H^+(aq)$	+ $OH^-(aq)$
Start	Some	0.10 $\underline{M}$	0.15 $\underline{M}$
Δ	+0.10	−0.10	−0.10
Finish	Some	0.00	0.05

$[H^+] = \dfrac{K_w}{[OH^-]} = \dfrac{1.0 \times 10^{-14}}{0.05} = 2 \times 10^{-13}$

$pH = -\log[H^+] = 12.7$

20. **(E)** Both reagents are 0.100 $\underline{M}$. At the equivalence point 50.0 mL of base has been added to 50.0 mL of acid.

$[HNO_2] = [OH^-] = \dfrac{50.0 \text{ mL} \times 0.100 \text{ mol L}^{-1}}{100. \text{ mL}} = 0.0500 \underline{M}$

The titration reaction is:
$HNO_2 + OH^-(aq) \Rightarrow NO^-(aq) + H_2O(l)$

The reaction has produced 0.0500 $\underline{M}$ $NO^-(aq)$, a base.
pH = 8.02 is the only base in the answer set.

A base reaction is the reverse of a neutralization.
$NO^-(aq) + H_2O(l) \Rightarrow HNO_2(aq) + OH^-(aq)$

	Finish	0.0500	Some	x	x

$K_b = \dfrac{K_w}{K_a} = \dfrac{1.0 \times 10^{-14}}{4.5 \times 10^{-4}} = 2.2 \times 10^{-11}$

$K_b = \dfrac{[HNO_2][OH^-]}{[NO^-]} = \dfrac{x^2}{0.0500} = 2.2 \times 10^{-11}$

$x = [OH^-] = 1.05 \times 10^{-6} \underline{M}$

$[H^+] = \dfrac{K_w}{[OH^-]} = \dfrac{1.00 \times 10^{-14}}{1.05 \times 10^{-6}} = 9.52 \times 10^{-9}$

$pH = -\log[H^+] = \mathbf{8.02}$

Set 7

Thermodynamics

1. A gas evolves 1420. J. of heat and as it is compressed from 30.0 L to 10.0 L by an opposing pressure of 2.00 atm.
   Note: 1.00 Liter-atm = 101 Joules.
   The internal energy change, ΔE, in <u>Joules</u> for the gas is:
   (A) +5460 J
   (B) −5460 J
   (C) +1420 J
   (D) −2620 J
   (E) +2620 J

2. In which of the following reactions would ΔH be most nearly equal to ΔE?
   (A) $2H_2(g) + O_2(g) \Rightarrow 2H_2O(l)$
   (B) $H_2(g) + Cl_2(g) \Rightarrow 2\ HCl(g)$
   (C) $C_2H_5Cl(g) \Rightarrow C_2H_4(g) + HCl(g)$
   (D) $2CO(g) + O_2(g) \Rightarrow 2CO_2(g)$
   (E) $C_2H_4(g) + 3O_2(g) \Rightarrow 2CO_2(g) + 2H_2O(l)$

3. Given that ΔH° is −350 kJ at 25°C for the process:

   $HCl(aq) + CaCO_3(s) \Rightarrow CaCl_2(aq) + H_2O(l) + CO_2(g)$

   In order to find ΔE° in kJ, which of the following numerical values could be substituted into the equation?

   ΔH° = ΔE° + ΔnRT
   (A) −298
   (B) 1.987 x 10⁻³
   (C) 0.0821
   (D) 1
   (E) 623

4. The combustion of 0.100 g of ethane (C_2H_6) caused the temperature of a bomb calorimeter and its contents to rise 1.74°C. The calorimeter and the water it contains has a heat capacity of 713 cal °C⁻¹. What is the energy change, in kcal mole⁻¹, for the combustion of ethane?
   (A) ΔE = −1240 kcal mole⁻¹
   (B) ΔH = −1240 kcal mole⁻¹
   (C) ΔE = − 372 kcal mole⁻¹
   (D) ΔH = − 372 kcal mole⁻¹
   (E) ΔE = − 123 kcal mole⁻¹

Multiple Choice Questions

5. Two (2.00) grams of hydrogen fluoride, HF (MW = 20.0), gas is bubbled into 400. mL of water at 20.6°C. All the gas dissolves, and the temperature of the solution rises to 24.3°C.

    Assume the density and specific heat (4.184 J g^{-1} °C^{-1}) of the solution is the same as that of pure water.
    What is the value of $\Delta H_{solution}$ for HF?
    (A) -62 kJ mole^{-1}
    (B) -41 kJ mole^{-1}
    (C) - 6.2 kJ mole^{-1}
    (D) + 6.2 kJ mole^{-1}
    (E) +62 kJ mole^{-1}

6. Find $\Delta H°$ for the reaction, in kcal.

    $$3V_2O_3(s) \Rightarrow V_2O_5 + 4VO(s)$$

    Standard heat of reactions, $\Delta H°$, are given in the table.

Reaction	$\Delta H°$, kcal
4VO(s) + O$_2$(g) $\Rightarrow$ 2V$_2$O$_3$(s)	-180
2V$_2$O$_3$(s) + O$_2$(g) $\Rightarrow$ 4VO$_2$(s)	-108
4VO$_2$(s) + O$_2$(g) $\Rightarrow$ 2V$_2$O$_5$(s)	- 58

    (A) +205 kcal   (D) -155 kcal
    (B) + 97 kcal   (E) -346 kcal
    (C) + 14 kcal

7. Given the thermochemical reaction data:

Reaction	$\Delta H°$, kJ
MnO$_2$(s) + CO(g) $\Rightarrow$ MnO(s) + CO$_2$(g)	-151
Mn$_3$O$_4$(s) + CO(g) $\Rightarrow$ 3MnO(s) + CO$_2$(g)	- 54
3Mn$_2$O$_3$(s) + CO(g) $\Rightarrow$ 2Mn$_3$O$_4$(s) + CO$_2$(g)	-142

    Find $\Delta H°$, in kJ, for the reaction:

    $$2MnO_2(s) + CO(g) \Rightarrow Mn_2O_3(s) + CO_2(g)$$

    (A) -385 kJ
    (B) -347 kJ
    (C) -313 kJ
    (D) -219 kJ
    (E) - 45 kJ

8. What is the heat of reaction, $\Delta H°$, in kJ mole^{-1} for the reaction?

$$CH_4(g) + 4Cl_2(g) \Rightarrow CCl_4(l) + 4HCl(g)$$

Substance	$\Delta H_f°$, kJ mole^{-1}
$CH_4(g)$	-74.9
$CCl_4(l)$	-135.6
$HCl(g)$	-92.5

(A) $\Delta H°$ cannot be computed without additional data.
(B) -153 kJ
(C) -303 kJ
(D) -430 kJ
(E) -580 kJ

9. The standard heat of reaction, $\Delta H°$, for the combustion reaction is -1985 kJ.

$$2(CH_3)_2CHOH(l) + 9O_2(g) \Rightarrow 6CO_2(g) + 8H_2O(l)$$

Substance	$\Delta H_f°$, kJ mole^{-1}
$CO_2(g)$	-394
$H_2O(l)$	-286

What is the standard heat of formation, $\Delta H_f°$, of isopropyl alcohol, $(CH_3)_2CHOH(l)$, in kJ mole^{-1}?
(A) +1310 kJ mole^{-1}
(B) - 678 kJ mole^{-1}
(C) -1330 kJ mole^{-1}
(D) -1990 kJ mole^{-1}
(E) -4650 kJ mole^{-1}

10. The $\Delta H_f°$ of methane, CH_4, is -17.9 kcal mole^{-1}.
$$C_{graphite} + 2H_2(g) \Rightarrow CH_4(g)$$

Bond	Bond Energy, kcal mole^{-1}
C–C bonds in graphite	171
H–H bonds in hydrogen	104

Calculate the bond energy of a C—H bond.
(A) 64 kcal mole^{-1}
(B) 73 kcal mole^{-1}
(C) 78 kcal mole^{-1}
(D) 90 kcal mole^{-1}
(E) 99 kcal mole^{-1}

11. The entropy of a perfect crystal is zero when the :
    (A) system is at equilibrium.
    (B) enthalpy change is zero.
    (C) entropy change is zero.
    (D) free energy change is zero.
    (E) system is at absolute zero.

12. Which reaction has the largest increase in entropy, ΔS?
    (A) $CO_2(s) \Leftrightarrow CO_2(g)$
    (B) $H_2(g) + Cl_2(g) \Leftrightarrow 2HCl(g)$
    (C) $C_{diamond} \Leftrightarrow C_{graphite}$
    (D) $2H_2(g) + O_2(g) \Leftrightarrow H_2O(g)$
    (E) $KNO_3(s) \Leftrightarrow KNO_3(l)$

13. The heat of fusion of sodium is 621 cal mole^{-1} at its normal melting temperature of 98.0 °C.

    $Na(s) \Leftrightarrow Na(l)$

    What is the entropy change, ΔS, in cal mol^{-1} K^{-1} at this temperature?
    (A) +371
    (B) +6.34
    (C) +1.67
    (D) $+1.58 \times 10^{-1}$
    (E) $+5.97 \times 10^{-1}$

14. For a spontaneous endothermic process:
    (A) $\Delta H < 0$
    (B) $S^o = 0$
    (C) $\Delta S > 0$
    (D) $\Delta G > 0$
    (E) $\Delta G = 0$

15. Which reaction will be spontaneous?

	ΔH, kJ	ΔS, J K^{-1}	T, K
(A)	+ 50	- 50	1000
(B)	+ 250	+ 400	300
(C)	+ 250	+ 50	300
(D)	- 100	- 400	250
(E)	- 100	- 400	200

16. A reaction has values of ΔH and ΔS which are both positive. The reaction:
    (A) cannot be spontaneous.
    (B) spontaneity is temperature dependent.
    (C) is exothermic.
    (D) free energy of is increasing.
    (E) must be at equilibrium.

Questions 17-18 deal with the reaction:

$$N_2O_5(g) \Leftrightarrow 2NO_2(g) + 1/2 O_2(g)$$

Substance	ΔH_f° kcal mol^{-1}	ΔG_f° kcal mol^{-1}	S° cal mol^{-1}
$N_2O_5(g)$	2.70	27.50	85.00
$NO_2(g)$	7.93	12.26	57.35
$O_2(g)$	----	----	49.00

17. What is the ΔG° of the reaction?
   (A) + 39.76 kcal
   (B) + 27.50 kcal
   (C) + 15.24 kcal
   (D) - 2.98 kcal
   (E) - 15.79 kcal

18. Use the equation:
    $$\Delta G = \Delta H - T\Delta S$$
    to calculate the value of ΔG, in kcal mol^{-1}, at 373 K.
   (A) - 7.06 kcal mol^{-1}
   (B) - 20.2 kcal mol^{-1}
   (C) - 3.01 kcal mol^{-1}
   (D) - 41.0 kcal mol^{-1}
   (E) + 13.1 kcal mol^{-1}

19. A gaseous reaction has a ΔG° of -24.1 kJ mol^{-1}.
    What is the approximate value of K_p at 298 K?
   (A) 3×10^8
   (B) 2×10^4
   (C) 1.0
   (D) 3×10^{-3}
   (E) 6×10^{-5}

20. The standard net cell potential, $\mathcal{E}^\circ$, is +1.50 volts for the reaction:
    $$Au^{3+}(aq) + 3e^- \Leftrightarrow Au(s)$$
    What is the standard free energy, ΔG°, in kJ?
   (A) -606
   (B) -434
   (C) -145
   (D) - 34.5
   (E)   0

Answer Key—Set 7

1.	(E)	6.	(B)	11.	(E)	16.	(B)
2.	(B)	7.	(D)	12.	(A)	17.	(D)
3.	(D)	8.	(D)	13.	(C)	18.	(A)
4.	(C)	9.	(C)	14.	(C)	19.	(B)
5.	(A)	10.	(E)	15.	(E)	20.	(D)

Explanations

1. **(E)** $\Delta E = q + w$
   $\Delta E = +1420 \text{ J} + (2 \text{ atm}) \times (20 \text{ L}) \times (101 \text{ J L}^{-1} \text{ atm}^{-1})$
   $\Delta E = +2620 \text{ J}$

2. **(B)** $\Delta H = \Delta E + \Delta nRT$
   $\Delta n = n(g)_{products} - n(g)_{reactants} = 0$ (for reaction (B))

3. **(D)** $\Delta H = \Delta E + \Delta nRT$
   $-350. = \Delta E + \dfrac{(1 \text{ mol}) \times (8.314 \text{ J mol}^{-1} \text{ K}^{-1}) \times (298 \text{ K})}{100 \text{ J kJ}^{-1}}$

4. **(C)**
   $q_v = (713 \text{ cal °C}^{-1}) \times (1.714 \text{ °C}) = -1240 \text{ cal}$
   $n = 0.100 \text{ g} \div 30.0 \text{ g mol}^{-1} = 0.00333 \text{ mol}$
   $\Delta E = -1240 \text{ cal} \div 0.00333 \text{ mol} = -372 \text{ kcal mol}^{-1}$

5. **(A)**
   $q_p = (4.184 \text{ J g}^{-1} \text{ °C}^{-1}) \times (400. \text{ g}) \times (24.3-20.6)$
   $q_p = -6200 \text{ J} = -6.2 \text{ kJ}$ (heat is evolved)
   $n_{HF} = 2.00 \text{ g} \div 20.0 \text{ g mol}^{-1} = 0.100 \text{ mol}$
   $\Delta H = 6.190 \text{ kJ} \div 0.100 \text{ mol} = 62 \text{ kJ mol}^{-1}$

6. **(B)** Use Hess' Law

Reaction	$\Delta H°$, kcal
$2V_2O_3(s) \Rightarrow 4VO(s) + O_2(g)$	+180
$V_2O_3(s) + \frac{1}{2} O_2(g) \Rightarrow 2VO_2(s)$	$\frac{1}{2} \times -108 = -54$
$2VO_2(s) + \frac{1}{2} O_2(g) \Rightarrow V_2O_5(s)$	$\frac{1}{2} \times -58 = -29$

$3V_2O_3(s) \Rightarrow V_2O_5 + 4VO(s)$ $\Delta H° = \Sigma \Delta H° = +97 \text{ kcal}$

7. **(D)** Use Hess' Law

Reaction	$\Delta H°$, kJ
$2MnO_2(s) + 2CO(g) \Rightarrow 2MnO(s) + 2CO_2(g)$	$2 \times -151 = -302$
$2MnO(s) + \frac{2}{3} CO_2(g) \Rightarrow \frac{2}{3} Mn_3O_4(s) + \frac{2}{3} CO(g)$	$\frac{2}{3} \times -54 = +36$
$\frac{2}{3} Mn_3O_4(s) + \frac{1}{3} CO_2(g) \Rightarrow Mn_2O_3(s) + \frac{1}{3} CO(g)$	$\frac{1}{3} \times -142 = +47$

$2MnO_2(s) + CO(g) \Rightarrow Mn_2O_3(s) + CO_2(g) \qquad \Delta H° = \Sigma \Delta H° = -219$

8. **(D)** $\Delta H° = \Sigma \Delta H_f°$ (Products) $- \Sigma \Delta H_f°$ (Reactants)

   $\Delta H° = ((-135.6) + (4 \times -92.5)) - (-74.9)$

   $\Delta H° = -430.$ kcal

9. **(C)** $\Delta H° = \Sigma \Delta H_f°$ (Products) $- \Sigma \Delta H_f°$ (Reactants)

   $\Delta H° = -1985 = ((6 \times -394) + (8 \times -286)) - (2x)$

   $x = \Delta H_f°(CH_3)_2CHOH = $ **-1330** kcal

10. **(E)** $\Delta H = $ (Bonds broken) + (Bonds formed)
    Bonds broken $= (+171) + (2 \times (+104))$
    Bonds broken $= +379$ kcal
    Bonds formed $= (4 \times (- kcal_{C-H}))$
    $\Delta H = -17.9$ kcal $= +379 + (4 \times (- kcal_{C-H}))$
    $kcal_{C-H} = $ **99** kcal mol^{-1}

11. **(E)** The third law of thermodynamics states that the entropy of a perfect crystal at absolute zero is 0.

12. **(A)** The change from (s) $\Rightarrow$ (g) will result in the highest increase in randomness or chaos.

13. **(C)** $\Delta G = \Delta H - T\Delta S$
    At fusion equilibrium, $\Delta G = 0$ and $\Delta H = T\Delta S$
    $(621 \text{ cal}) = (371 \text{ K}) \times (\Delta S)$
    $\Delta S = 1.67$ cal K^{-1}

14. **(C)** When a reaction is endothermic, $\Delta H > 0$. For the reaction to be spontaneous, ΔG must be negative ($\Delta G < 0$) and ΔS must be positive ($\Delta S > 0$)

15. **(E)**
$$\Delta G = \Delta H - \frac{T\Delta S}{1000}$$
$$\Delta G = (-100 \text{ kJ}) - \frac{(200 \text{ K}) \times (-400 \text{ J K}^{-1})}{1000 \text{ J kcal}^{-1}} = -20 \text{ kJ}$$

When $\Delta G < 0$, the reaction is spontaneous.

16. **(B)** At high temperature, $T\Delta S$ can exceed ΔH to $\Delta G < 0$ for a spontaneous reaction.

17. **(D)**
$$\Delta G° = \Sigma \Delta G_f° \text{ (Products)} - \Sigma \Delta G_f° \text{ (Reactants)}$$
$$\Delta G° = ((2 \times 12.26) + (0)) - (27.50))$$
$$\Delta G° = -2.98 \text{ kcal}$$

18. **(A)**
$$\Delta H = \Sigma \Delta H_f \text{ (Products)} - \Sigma \Delta H_f \text{ (Reactants)}$$
$$\Delta H = ((2 \times 7.93) + (0)) - (2.70)$$
$$\Delta H = +13.16 \text{ kcal}$$

$$\Delta S = \Sigma S_f \text{ (Products)} - \Sigma S_f \text{ (Reactants)}$$
$$\Delta S = ((2 \times 57.35) + (½ \times 49.00)) - (85.00)$$
$$\Delta S = +54.2 \text{ cal K}^{-1}$$

$$\Delta G = \Delta H - \frac{T\Delta S}{1000}$$
$$\Delta G = (+13.16 \text{ kcal}) - \frac{(373 \text{ K}) \times (54.2 \text{ cal K}^{-1})}{1000 \text{ cal kcal}^{-1}}$$
$$\Delta G = -7.06 \text{ kcal}$$

19. **(B)**
$$\Delta G° = -RT \ln K$$
$$(-24{,}100 \text{ J mol}^{-1}) = -(8.31 \text{ J mol}^{-1} \text{ K}^{-1}) \times (298 \text{ K}) \times \ln K$$
$$\ln K = 9.72$$
$$K = e^{9.72} \text{ (inverse ln K)} = 2 \times 10^4$$

20. **(D)**
$$\Delta G° = -nF\mathcal{E}°$$
$$\Delta G° = -(3 \text{ mol}) \times (96{,}500 \text{ cal mol}^{-1} \text{ volt}^{-1}) \times (1.50 \text{ volt})$$
$$\Delta G° = -434{,}000 \text{ cal} = -434 \text{ kcal}$$

Set 8

Electrochemistry

Note: Some of the problems require that you consult a table of 'Standard Reduction Potentials'. A table is found in Chapter 8 of Part I, "Review of Selected Topics".

1. Which equation correctly represents the change from ClO_3^- to $HClO_2$ in acidic solution?
   - (A) $ClO_3^- + 3H^+ \Rightarrow HClO_2 + H_2O + 2e^-$
   - (B) $ClO_3^- + 2H_2O + 2e^- \Rightarrow HClO_2 + 3OH^-$
   - (C) $ClO_3^- + 2H_2O + 2e^- \Rightarrow HClO_2 + 2OH^-$
   - (D) $ClO_3^- + 2H_2O \Rightarrow HClO_2 + 3OH^- + 2e^-$
   - (E) $ClO_3^- + 3H^+ + 2e^- \Rightarrow HClO_2 + H_2O$

2. Which balanced equation represents the change from $S_2O_3^{2-}$ to SO_3^{2-} in basic solution?
   - (A) $S_2O_3^{2-} + 6OH^- \Rightarrow 2SO_3^{2-} + 3H_2O + 4e^-$
   - (B) $S_2O_3^{2-} + 6OH^- + 4e^- \Rightarrow 2SO_3^{2-} + 3H_2O$
   - (C) $S_2O_3^{2-} + 2e^- \Rightarrow 2SO_3^{2-}$
   - (D) $S_2O_3^{2-} + 3H_2O + 4e^- \Rightarrow 2SO_3^{2-} + 6H^+$
   - (E) $S_2O_3^{2-} + 3H_2O \Rightarrow 2SO_3^{2-} + 6H^+ + 4e^-$

3. Which is the correct oxidation half–reaction for the reaction between Na_2O_2 and $HSnO_2^-$ to give $HSnO_3^-$ and OH^- in basic solution?
   - (A) $Na_2O_2 + 2H_2O + 2e^- \Rightarrow 2Na^+ + 4OH^-$
   - (B) $Na_2O_2 + 2H_2O \Rightarrow 2Na^+ + 4OH^- + 2e^-$
   - (C) $Na_2O_2 + HSnO_2^- + H_2O \Rightarrow HSnO_3^- + 2Na^+ + 2OH^-$
   - (D) $HSnO_2^- + 2OH^- + 2e^- \Rightarrow HSnO_3^- + H_2O$
   - (E) $HSnO_2^- + 2OH^- \Rightarrow HSnO_3^- + H_2O + 2e^-$

4. This equation is correctly balanced in basic solution. Which is the oxidizing agent for the reaction as written?
   $$8Al + 5OH^- + 3NO_3^- + 18H_2O \Rightarrow 8Al(OH)_4^- + 3NH_3$$
   - (A) Al
   - (B) OH^-
   - (C) NO_3^-
   - (D) $Al(OH)_4^-$
   - (E) NH_3

5. Which can be reduced by good oxidizing agents and oxidized by good reducing agents?
   - (A) NH_3
   - (B) NO_3^-
   - (C) I^-
   - (D) SO_3^{2-}
   - (E) MnO_4^-

138   Multiple Choice Questions

6. An 1.0 M aqueous solution of sodium nitrate, NaNO$_3$(aq), is electrolyzed using inert electrodes. What is the substance formed at the cathode?
    (A) Na
    (B) N$_2$
    (C) NO$_2$
    (D) H$_2$
    (E) O$_2$

7. An 1.0 M aqueous solution of zinc iodide, ZnI$_2$(aq), is electrolyzed using inert electrodes. What is the substance formed at the anode?
    (A) Zn
    (B) I$_2$
    (C) H$_2$O
    (D) H$_2$
    (E) O$_2$

8. The number of Faradays required to produce 9 g of aluminum, Al, by the electrolysis of molten aluminum oxide, Al$_2$O$_3$, is:
    (A) 9 Faradays
    (B) 4 Faradays
    (C) 3 Faradays
    (D) 2 Faradays
    (E) 1 Faraday

9. A current of 3.25 amperes is used to electrolyze a solution of copper II sulfate, CuSO$_4$. How many hours will it take to deposit 12.71 grams of copper, Cu?
    (A) 6.60 hours
    (B) 3.30 hours
    (C) 1.65 hours
    (D) 0.400 hours
    (E) 0.200 hours

10. Two cells, one containing gold nitrate, Au(NO$_3$)$_3$, and the other copper II sulfate, CuSO$_4$, are connected in series. During the electrolysis, 98.5 grams of gold are deposited in one cell. What weight of copper is deposited in the second cell?
    (A) 21.2 g
    (B) 31.8 g
    (C) 47.6 g
    (D) 63.5 g
    (E) 98.5 g

11. A voltaic cell is set up using the system:
    Fe/Fe^{2+} || Ag$^+$/Ag
    The cathode reaction produces:
    (A) Fe
    (B) Fe^{2+}
    (C) Ag$^+$
    (D) Ag
    (E) H$_2$

12. A voltaic cell is set up using the system:
    Zn/Zn^{2+} || Au^{3+}/Au
    Which is true about movement through the salt bridge?
    (A) Zinc ions, Zn^{2+}, move towards the cathode.
    (B) Zinc ions, Zn^{2+}, move towards the anode.
    (C) Electrons, e$^-$, move towards the cathode.
    (D) Gold ions, Au^{3+}, move toward the anode.
    (E) Negative 'spectator' ions move towards the anode.

13. A voltaic cell is set up using the system:
    Ni/Ni^{2+} || Cu^{2+}/Cu
    Which reaction would occur at the anode?
    (A) Ni $\Rightarrow$ Ni^{2+} + 2e$^-$
    (B) Cu $\Rightarrow$ Cu^{2+} + 2e$^-$
    (C) Ni^{2+} + 2e$^-$ $\Rightarrow$ Ni
    (D) Cu^{2+} + 2e$^-$ $\Rightarrow$ Cu
    (E) Both (B) and (C)

14. A voltaic cell employs the reaction:
    Sn(s) + 2Ag$^+$(aq) $\Rightarrow$ Sn^{2+}(aq) + 2Ag(s)
    The voltage produced by this reaction under standard state condition at 25°C. is:
    (A) 1.74 volts
    (B) 1.46 volts
    (C) 0.94 volts
    (D) 0.66 volts
    (E) 0.52 volts

15. The standard EMF, $\mathcal{E}°$, is +0.68 volts for the voltaic cell:
    In/In^{3+} || Cu^{2+}/Cu
    The standard oxidation potential for the reaction:
    In $\Rightarrow$ In^{3+} + 3e$^-$
    (A) is +0.34 volts.
    (B) is -0.34 volts.
    (C) is +1.02 volts.
    (D) is -1.02 volts.
    (E) is 0.00 volts.

16. From a table of standard reduction potentials it can be concluded that:
    (A) Zn^{2+} reacts spontaneously with H$_2$.
    (B) Ag reacts spontaneously with 1 M hydrochloric acid, HCl.
    (C) Ag reacts spontaneously with 1 M Zn^{2+}.
    (D) Ag reacts spontaneously with 1 M nitric acid, HNO$_3$.
    (E) Zn^{2+} will liberate H$_2$ from 1 M HCl.

17. From the table of standard reduction potentials find a species that will convert Cu^{2+} to Cu but will not convert Fe^{2+} to Fe.
    (A) Ag
    (B) H$_2$
    (C) Sn^{4+}
    (D) I$^-$
    (E) Zn

18. Which is the strongest reducing agent?
    (A) Ag
    (B) Ag$^+$
    (C) Fe
    (D) Fe^{2+}
    (E) H$_2$

19. The potential of the cell:
    Cd/Cd^{2+} || Ni^{2+}/Ni
    would be made more positive by increasing the:
    (A) size of the Cd electrode.
    (B) size of the Ni electrode.
    (C) volume of the electrolytes.
    (D) concentration of Cd^{2+}.
    (E) concentration of Ni^{2+}.

20. A voltaic cell has an $\mathcal{E}°$ of 1.56 volts and the overall reaction:
    $$Zn(s) + 2Ag^+(aq) \Leftrightarrow Zn^{2+}(aq) + 2Ag(s)$$

    The [Zn^{2+}] is 0.00010 M and the [Ag$^+$] is 0.10 M.

    What is the voltage, $\mathcal{E}$, of this cell?
    (A) 1.65 volts.   (D) 1.50 volts.
    (B) 1.62 volts.   (E) 1.47 volts.
    (C) 1.56 volts.

Electrochemistry

Answer Key—Set 8

1.	(E)	6.	(D)	11.	(D)	16.	(D)
2.	(A)	7.	(B)	12.	(A)	17.	(B)
3.	(E)	8.	(E)	13.	(A)	18.	(C)
4.	(C)	9.	(B)	14.	(C)	19.	(E)
5.	(D)	10.	(C)	15.	(A)	20.	(B)

Explanations

1. **(E)**

   BALANCE O: $ClO_3^- \Rightarrow HClO_2 + H_2O$
   BALANCE H: $ClO_3^- + 3H^+ \Rightarrow HClO_2 + H_2O$
   BALANCE CHARGE: $ClO_3^- + 3H^+ + 2e^- \Rightarrow HClO_2 + H_2O$

2. **(A)**

   BALANCE S: $S_2O_3^{2-} \Rightarrow 2SO_3^{2-}$
   BALANCE O: $S_2O_3^{2-} + 6OH^- \Rightarrow 2SO_3^{2-} + 3H_2O$
   BALANCE CHARGE: $S_2O_3^{2-} + 6OH^- \Rightarrow 2SO_3^{2-} + 3H_2O + 4e^-$

3. **(E)** Oxidation is the loss of electrons.
   When the 2 half–reactions are balanced:
   OXIDATION: $HSnO_2^- + 2OH^- \Rightarrow HSnO_3^- + 4H_2O + 2e^-$
   REDUCTION: $Na_2O_2 + 2H_2O + 2e^- \Rightarrow 2Na^+ + 4OH^-$

4. **(C)** The oxidizing agent is reduced in the reaction.
   METHOD 1: The oxidation state of nitrogen is being reduced from +5 in NO_3^- to -3 in NH_3.
   METHOD 2: Balance the half-reactions to see which is the reduction.
   $NO_3^- + 6H_2O + 8e^- \Rightarrow NH_3 + 9OH^-$

5. **(D)** The oxidation number of sulfur in the sulfite ion, SO_3^{2-}, is +4, an intermediate value for sulfur. Strong oxidizers can oxidize sulfite ion to sulfate ion, SO_4^{2-}. Sulfite can be reduced to either sulfur or to sulfide ion, S^{2-}, by strong reducing agents.

6. **(D)** Na^+ and NO_3^- do not undergo electrolysis in aqueous solution.
   Reduction occurs at the cathode (*Red Cat*).
   REDUCTION: $2H_2O \Rightarrow O_2 + 4H^+ + 4e^-$
   OXIDATION: $2H_2O + 2e^- \Rightarrow H_2 + 2OH^-$

7. **(B)** Zn^{2+} and I^- are electrolyzed better than H_2O.
   OXIDATION: (ANODE) $2I^- \Rightarrow I_2 + 2e^-$
   REDUCTION: (CATHODE) $Zn^{2+} + 2e^- \Rightarrow Zn$

142 Multiple Choice Questions

8. **(E)** $n_{Al} = 9\ g \div 27\ g\ mol^{-1} = \frac{1}{3}\ mol$
$Al^{3+} + 3e^- \Rightarrow Al$
Faradays $= n_{e^-} = 3n_{Al} = 3 \times \frac{1}{3} =$ **1 F**

9. **(B)** $n_{Cu} = 12.71\ g \div 63.55\ g\ mol^{-1} = 0.2000\ mol$
$Cu^{2+} + 2e^- \Rightarrow Cu$
$n_{e^-} = 2n_{Cu} = 0.4000\ mol$
Time $= \dfrac{0.4000\ mol\ e^- \times 96{,}500\ coul\ mol^{-1}}{3.25\ coul\ sec^{-1} \times 3600\ sec\ hr^{-1}} =$ **3.30 hr**

10. **(C)** $n_{Au} = 98.5\ g \div 197\ g\ mol^{-1} = 0.500\ mol$
$Au^{3+} + 3e^- \Rightarrow Au$
$n_{e^-} = 3n_{Au} = 1.50\ mol$
$Cu^{2+} + 2e^- \Rightarrow Cu$
$n_{Cu} = \frac{1}{2}n_{e^-} = 0.750\ mol$
$wt_{Cu} = 0.750\ mol \times 63.55\ g\ mol^{-1} =$ **47.6 g**

11. **(D)** Reduction occurs at the cathode (*Red Cat*).
CATHODE: $Ag^+ + e^- \Rightarrow$ **Ag**

12. **(A)** CATHODE: $Au^{3+} + 3e^- \Rightarrow Au$
The $[Au^{3+}]$ decreases in the cathode half–reaction.
There must be a migration of (+) ions (Zn^{2+}) to the cathode and (−) ions to the anode to balance the charge in each cell.
Electrons can only move in the external circuit.

13. **(A)** **Oxidation occurs at the anode** (*An Ox*).
ANODE: $Ni \Rightarrow Ni^{2+} + 2e^-$

14. **(C)**

			$\mathcal{E}°$
ANODE:	Sn	$\Rightarrow Sn^{2+} + 2e^-$	+0.14 volts
CATHODE:	$2Ag^+ + 2e^-$	$\Rightarrow 2Ag$	+0.80 volts
			+0.94 volts

Electrochemistry   143

15. **(A)**

			$\mathcal{E}°$
ANODE:	In	$\Rightarrow$ In^{3+} + 3e$^-$	$\mathcal{E}°_{In}$
CATHODE:	Cu^{2+} + 2e$^-$	$\Rightarrow$ Cu	+0.34 volts
			+0.68 volts

$\mathcal{E}°_{In}$ = 0.34 volts

16. **(D)** A **positive** net cell potential is **spontaneous**. Nitric acid is a strong oxidizing agent.

OXIDATION: 3Ag $\Rightarrow$ 3Ag$^+$ + 3e$^-$ −0.80 volts
REDUCTION: NO$_3^-$ + 4H$^+$ + 3e$^-$ $\Rightarrow$ NO + 2H$_2$O   +0.96 volts
                                                                                        +0.16 volts

17. **(B)** H$_2$ is between Fe^{2+} and Cu^{2+} in the table. It will give a positive net cell potential with Cu^{2+}, but not Fe^{2+}.

Fe^{2+} + 2e$^-$ $\Rightarrow$ Fe   −0.45 volts
2H$^+$ + 2e$^-$ $\Rightarrow$ H$_2$ +0.00 volts
Cu^{2+} + 2e$^-$ $\Rightarrow$ Cu   +0.34 volts

18. **(C)** The strongest reducing agent is the most easily oxidized. The highest oxidation potential of those given is that of **iron, Fe**, whose $\mathcal{E}°_{oxid}$ is +0.45 volts.

19. **(E)** The net cell reaction is:
    Cd + Ni^{2+} $\Leftrightarrow$ Cd^{2+} + Ni

Increasing [Ni^{2+}] or decreasing [Cd^{2+}] will favor the forward reaction and the net cell potential.
Changing electrode size or electrolyte volume will have no effect on the voltage. The amount of a solid or a liquid does not effect an equilibrium reaction.

20. **(B)** $\mathcal{E} = \mathcal{E}° - \dfrac{0.059}{n} \log \dfrac{[Zn^{2+}]}{[Ag^+]^2}$

$\mathcal{E} = 1.56 - \dfrac{0.059}{2} \log \dfrac{[0.00010]}{[0.10]^2}$

$\mathcal{E} = 1.62$ volts

Part III

Free Response Section

About Section II

Free Response Problems

Section II of the examination generally constitutes 55% of the final score. It is allotted between 75 and 90 minutes with 90 minutes being the most usual time allowance. This section is a distinctive and important feature of the examination.

Parts A and B are problems that permit the candidate to demonstrate an ability to apply chemical principles to the solution of college–level problems. Part A is worth 25 points and has always been a question about 'Equilibrium'. Part B is also worth 25 points and permits the choice of one problem (from two given) dealing with 'Electrochemistry', 'Kinetics', 'Stoichiometry' or 'Thermodynamics'. No more than 20–25 minutes should be spent on each part.

The table is a history of the types of problems given for each of the past years. It should be understood that no prediction can be made for any specific test. Section II questions are released each year. Collections of Section II questions, answers and explanations are published by the Educational Testing Service.

Year	Part A *Equilibrium*	Part B (Choose 1 of 2 questions)	
1976	*Acid-Base* [H_3O^+], K_a, K_{sp}	*Electrochemistry* $\Delta G°$, $E°$, Faraday	*Stoichiometry* Molality, Raoult, Colligative Properties
1977	*Acid-Base* [H_3O^+], buffers, pH	*Kinetics* K_{eq}, Rate	*Thermo* $\Delta G°$, $\Delta S°$, K_{eq}
1978	*Acid-Base* Titration	*Electrochemistry* $\Delta G°$, $E°$, Stoic Stoichiometry	*Thermo* $\Delta H°$, K, LeChatelier,
1979	*Solubility* K_{sp}	*Stoichiometry* Solutions, molarity	*Thermo* $\Delta G°$, $\Delta S°$
1980	*Acid-Base* [H_3O^+], K_b, K_{sp}	*Electrochemistry* $E°$, Nernst	*Stoichiometry* Colligative, empirical, gas laws
1981	*Gaseous* K_p, gas law	*Kinetics* Rate and k	*Stoichiometry* Balance, titration
1982	*Acid-Base* [H_3O^+], pH, buffers	*Electrochemistry* Electrolysis, Faraday	*Stoichiometry* Empirical, percent composition

Year	Part A *Equilibrium*	Part B (Choose 1 of 2 questions)	
1983	*Gaseous* Law, Dalton K_p, K_c	*Stoichiometry* Solutions, titration molar mass	*Thermo* $\Delta G°$, K, $\Delta H°$, $\Delta S°$
1984	*Acid-Base* [OH⁻], K_b, K_a, solubility	*Kinetics* Rate law, k, mechanisms	*Thermo* $\Delta G°$, $\Delta H°$, $\Delta S°$
1985	*Solubility* Selective precipitation	*Electrochemistry* Balance, Faraday, $E°$, $\Delta G°$	*Stoichiometry* Empirical, molecular, colligative, MW
1986	*Acid-Base* K_a, pH, buffers	*Electrochemistry* Faraday, balance, redox, titration	*Stoichiometry* Gas law, molar mass, empirical
1987	*Acid-Base* [OH⁻], pH, buffers, K_{sp}	*Kinetics* Rate law, stoich	*Stoichiometry* Solutions, titration, percent composition
1988	*Gaseous* Gas law, K_c, K_p	*Electrochemistry* Electrochemical cells, K	*Thermo* Hess' Law, $\Delta G°$, $\Delta H°$, $\Delta S°$, K
1989	*Acid-Base* MW, K_a, buffers, pH	*Electrochemistry* Titration, stoich., redox reaction, Faraday	*Thermo* $\Delta G°$, $\Delta H°$, $\Delta S°$, T, vapor pressure

The Part A and B problems usually have three or four sub-parts. Sub-parts (a) and (b) are standard questions that have usually been practiced previously. Sub-parts (c) and (d) are a little more elegant (difficult) since they seem to be unique. Proper application of basic equilibrium principles should yield a solutions to these sub-parts.

The correct strategy for candidates working on this section is to make sure to:

1. show a logical series of steps that can lead to an answer.

2. acknowledge that only one-third credit is given for a right answer without supporting steps.

3. appreciate that partial credit is given for each correct step in the attempt to answer the question, even if the question is not fully answered.

4. realize that some problems have several valid approaches to the correct answer.

5. make use of the fact that an incorrect answer from one part will not be penalized when used in other parts of the question.

6. consistently use units in equations and in labeling answers.

7. use significant figures correctly.

Chapter 1

Part A Sample Problems

1. **Gaseous Equilibrium**

    Solid ammonium chloride is placed in an evacuated vessel and heated until it decomposes.
    $$NH_4Cl(s) \Leftrightarrow NH_3(g) + HCl(g)$$

    (a) Equilibrium is reached at a temperature of 202°C, and the total pressure inside the vessel is 4.40 atm.

    What is the numerical value of K_p at 202°C?

    (b) Ammonia gas, $NH_3(g)$, is added to the vessel, which is kept at 202°C. The system returns to equilibrium, and the partial pressure of the NH_3 is found to be three-times that of the HCl.

    What are the partial pressures of NH_3 and HCl?

    (c) To an empty 1.00–Liter vessel is added 0.130 mole each of NH_3 and HCl at 202°C. How many moles of solid NH_4Cl will be formed at this temperature?

    (d) Explain the effect on the equilibrium constant of raising the temperture to 600 K.

2. **Solubility Equilibrium**

    A saturated solution of magnesium hydroxide, $Mg(OH)_2$, has a magnesium ion concentration of 1.65×10^{-4} M at 25°C.

    (a) What is the value of the solubility product constant, K_{sp}, at 25°C.

    (b) What is the molar solubility of $Mg(OH)_2$ in 0.10 M $Mg(NO_3)_2$ solution at 25°C.

    (c) To 350.–mL of a 0.150 M $Mg(NO_3)_2$ solution, 150.–mL of a 0.500 M NaOH solution is added. What are the $[Mg^{2+}]$ and $[OH^-]$ in the resulting solution at 25°C?

150 Free Response Section

3. **Acid-Base Equilibrium**

   In water, acetylsalicylic acid (aspirin), $HC_9H_7O_4$, is a weak acid that has an equilibrium constant, K_a, equal to 2.75×10^{-5} at 25°C. A 0.400-Liter sample of a 0.100 M solution of the acid is prepared.

   (a) What are the equilibrium concentrations of $C_9H_7O_4^-$, H_3O^+, OH^- and what is the pH of the solution at 25°C?

   (b) What is the K_b for the reaction:
   $C_9H_7O_4^- + H_2O \Leftrightarrow HC_9H_7O_4 + OH^-$

   (c) To 0.2000-Liter of the solution, 3.03 grams of sodium acetysalicylate, $NaC_9H_7O_4$ (202 g mol^{-1}), is added. The salt dissolves completely and the volume of the solution remains unchanged.
   Calculate the pH of the resulting solution at 25°C.

   (d) To the remaining 0.200-Liter of the original solution, 0.100-Liter of 0.100 M NaOH solution is added.
   Calculate the [OH$^-$] for the resulting solution at 25°C.

Sample Answers and Explanations

1. **Gaseous Equilibrium**

   (a) Answer: K_p = 4.84 atm^2

	$NH_4Cl(s)$ $\Leftrightarrow$	$NH_3(g)$ +	$HCl(g)$
Equilibrium Partial Pressure	0	x	x

   P_{total} = $0 + x + x$ = 4.40 atm.
   x = P_{NH3} = P_{HCl} = 2.20 atm.
   K_p = $(P_{NH3}) \times (P_{HCl})$ = $(2.20 \text{ atm})^2$ = 4.84 atm^2

   (b) Answers: P_{HCl} = 1.27 atm
                 P_{NH3} = 3.81 atm

   K_p = $(P_{NH3}) \times (P_{HCl})$ = $(3P_{HCl}) \times (P_{HCl})$ = 4.84 atm^2
   P_{HCl} = 1.27 atm
   P_{NH3} = $3P_{HCl}$ = 3×1.27 = 3.81 atm

(c) Answer: $n_{NH_4Cl} = 0.0736$ mol

Since moles in a 1.00-L volume are given (molarity), K_c should be handier to use than K_p.

$K_c = K_p(1/RT)^{\Delta n}$
$K_c = 4.84$ atm^2 × (1 ÷ (0.0821 L-atm mol^{-1} K^{-1} × 475 K))^2
$K_c = 0.00318$ M^2

	NH$_4$Cl(s) ⇔	NH$_3$(g) +	HCl(g)
Start	0	0.130	0.130
Δ	+x	-x	-x
Equilibrium	x	0.130 - x	0.130 - x

K_c = [NH$_3$][HCl] = [0.130 - x]^2 = 0.00318 M^2
x = n_{NH_4Cl} = 0.0736 mol

(d) The equilibrium constant will be larger at 600 K.

The forward reaction is endothermic. An increase in temperature favors the endothermic (heat consuming) reaction leading to an increase in the partial pressure of the products.

2. Solubility Equilibrium
   (a) Answer: $K_{sp} = 1.80 \times 10^{-11}$

   [Mg^{2+}] = 1.65×10^{-4} M
   [OH$^-$] = 2 × [Mg^{2+}] = 3.30×10^{-4} M
   K_{sp} = [Mg^{2+}][OH$^-$]^2 = [1.65×10^{-4}][3.30×10^{-4}]^2
   K_{sp} = 1.80×10^{-11}

   (b) **Answer: x = molar solubility = 6.71×10^{-6} M**

   Let x equal the molar solubility of Mg(OH)$_2$.

	Mg(OH)$_2$(s) ⇔	Mg^{2+}(aq) +	2OH$^-$(aq)
Start 0	0.10 M		0 M
Δ	-x	+x	+2x
Finish	Some - x	0.10 + x	2x

   Note: x is small compared to 0.10, and is ignored.

   K_{sp} = [0.10][2x]^2 = 1.80×10^{-11}
   x = molar solubility = 6.71×10^{-6} M

(c) Answers: $[Mg^{2+}] = 0.030 \underline{M}$
$[OH^-] = x = 2.45 \times 10^{-5} \underline{M}$

$[Mg^{2+}] = \dfrac{350. \text{ mL} \times 0.150 \text{ mol L}^{-1}}{500. \text{ mL}} = 0.105 \underline{M}$

$[OH^-] = \dfrac{150. \text{ mL} \times 0.500 \text{ mol L}^{-1}}{500. \text{ mL}} = 0.150 \underline{M}$

Let x = amount of OH⁻ formed as a new equilibrium is established ('bounce back').

	$Mg(OH)_2(s)$ ⇔	$Mg^{2+}(aq)$ +	$2OH^-(aq)$
Start	0	0.105 $\underline{M}$	0.150 $\underline{M}$
Δ₁	+0.120	−0.075	−0.150
Precip.	Some	0.030	0
Δ₂	−x	+½	+x
Finish	Some − x	0.030 + ½	x

Note: ½ is small compared to 0.030, and is ignored.

$K_{sp} = [Mg^{2+}][OH^-]^2 = [0.030][x]^2 = 1.80 \times 10^{-11}$
$[Mg^{2+}] = 0.030 \underline{M}$
$[OH^-] = x = 2.45 \times 10^{-5} \underline{M}$

3. Acid-Base Equilibrium

(a) Answers: $[H_3O^+]$ and $[C_9H_7O_4^-] = 1.66 \times 10^{-3} \underline{M}$
pH = 2.780; $[OH^-] = 6.02 \times 10^{-12} \underline{M}$

	$HC_9H_7O_4(aq)$ ⇔	$H^+(aq)$ +	$C_9H_7O_4^-(aq)$
Start	0.100 $\underline{M}$	0.0 $\underline{M}$	0.0 $\underline{M}$
Δ	−x	+x	+x
Finish	0.100 − x	x	x

K_a = $\dfrac{[H^+][C_9H_7O_4^-]}{[HC_9H_7O_4]}$ = $\dfrac{[x][x]}{[0.100]}$ = 2.75×10^{-5}

x = $[H^+(aq)]$ = $[H_3O^+]$ = $[C_9H_7O_4^-]$ = $1.66 \times 10^{-3} \underline{M}$

pH = $-\log[H^+]$ = 2.780

K_w = $[H^+][OH^-]$ = $[1.66 \times 10^{-3}][x]$ = 1.0×10^{-14}

x = $[OH^-]$ = $6.02 \times 10^{-12} \underline{M}$

(b) Answer: $K_b = 3.6 \times 10^{-10}$

$K_a \times K_b$ = K_w = 1.0×10^{-14}
$(2.75 \times 10^{-5}) \times K_b$ = 1.0×10^{-14}
K_b = 3.6×10^{-10}

(c) Answer: pH = 4.436

$n_{C_9H_7O_4^-}$ = 3.03 g ÷ 202 g mol⁻¹ = 0.0150 mol
molarity = 0.0150 mol ÷ 0.200 L = 0.0750 $\underline{M}$
pH = $pK_a + \log \dfrac{[C_9H_7O_4^-]}{[HC_9H_7O_4]}$ = $4.561 + \log \dfrac{[0.0750]}{[0.100]}$
pH = 4.436

(d)   Answer: [OH⁻] = 3.6×10^{-10} M

$$[OH^-] = \frac{0.100\text{-L} \times 0.100 \text{ mol L}^{-1}}{0.300\text{-L}} = 0.0333 \text{ M}$$

$$[HC_9H_7O_4] = \frac{0.200\text{-L} \times 0.100 \text{ mol L}^{-1}}{0.300\text{-L}} = 0.0667 \text{ M}$$

	$C_9H_7O_4^-$(aq) +	H_2O	⇌ $HC_9H_7O_4$(aq) +	OH^-(aq)
Start	0	Some	0.0667 M	0.0333 M
Δ₁	+0.0333	+0.0333	−0.0333	−0.0333
Finish₁	0.0333	Some	0.0334	0.00
Δ₂	−x	−x	+x	+x
Finish₂	0.0333 − x	Some	0.0334 + x	x

$$K_b = \frac{[HC_9H_7O_4][OH^-]}{[C_9H_7O_4^-]} = \frac{[0.0334]x}{[0.0333]} = 3.6 \times 10^{-10}$$

$x = [OH^-] = 3.6 \times 10^{-10}$ M

Chapter 2

Part B Sample Problems

1. **Electrochemistry: Electrolytic Cells**

    An acidic solution contains lead(II) ions, Pb^{2+}, and an anion that is not electrolyzed. Electrolysis of the solution between lead electrodes produces lead IV oxide, PbO_2, at the anode and lead, Pb, at the cathode.

    (a) Write the balanced equation for the anode reaction.

    (b) A current of 1.75 amperes is applied to the system for 20.0 minutes. How many grams of Pb will be plated out at the cathode?

    (c) In series with the cell described above is another cell consisting of a solution of chromium(III) nitrate, $Cr(NO_3)_3$, and two platinum electrodes. How many grams of chromium, Cr, will be plated out when 49.5 grams of lead, Pb, are produced in the other cell?

2. **Electrochemistry: Voltaic Cells**

    A voltaic cell consists of a copper electrode in a solution of copper(II) ions, Cu^{2+}(aq), and a paladium electrode in a solution of paladium(II) ions, Pd^{2+}(aq).
    The copper electrode is the anode. The reduction potential for copper II ions is +0.342 volts.

    (a) Write the half–cell equation for the reaction that occurs at the cathode.

    (b) The standard cell voltage, $\mathcal{E}°$, is 0.609 volts. What is the reduction potential for the Pd^{2+}/Pd half–reaction?

    (c) What is the value of the equilibrium constant, K, for this reaction at 25°C?

    (d) A non–standard cell consisting of $Cu/Cu^{2+} \mid\mid Pd^{2+}/Pd$. The molarity of the copper(II) ion, Cu^{2+}, is 3.00 M and the molarity of the paladium (II) ion, $[Pd^{2+}]$, is 0.0500 M. What is the cell voltage?

3. **Kinetics**

   The data were obtained for the initial rate for the reaction:
   $$2NO(g) + O_2(g) \Rightarrow 2NO_2(g)$$

Data	[NO]	[O$_2$]	Initial rate; mol L^{-1} sec^{-1}
Run 1	0.20 M	0.20 M	4.64 x 10^{-8}
Run 2	0.20	0.40	9.28 x 10^{-8}
Run 3	0.40	0.40	3.71 x 10^{-7}
Run 4	0.10	0.10	5.80 x 10^{-9}

   (a) What is the rate law equation for the reaction?

   (b) What is the value of the specific rate law constant, including units?

   (c) The following mechanism has been suggested:

   1. $NO(g) + O_2(g) \Leftrightarrow NO_3(g)$ (Equilibrium)
   2. $NO_3(g) + NO(g) \Rightarrow 2NO_2(g)$ (Slow)

   Show that this mechanism leads to the observed rate equation.

   (d) The reaction rate is increased five-fold (5 times) when the temperature is increased from 1400 K to 1500 K. What is the energy of activation for the reaction?

4. **Stoichiometry: Empirical Formulas and Colligative Properties**

   Ascorbic acid (Vitamin C) is a compound known to contain carbon, hydrogen and oxygen.

   (a) Combustion analysis of a 1.000 gram sample of ascorbic acid yields 1.500 g of CO_2 and 0.409 g of H_2O. What is the empirical formula of this acid?

   (b) When another 1.00 g sample of ascorbic acid is dissolved in 10.0 g of water the freezing point of the solution is -1.06°C. At this temperature the dissociation of ascorbic acid is negligible. The freezing point constant, K_f, of water is 1.86 °C molal^{-1}. What is the molecular weight of ascorbic acid?

   (c) What is the molecular formula of ascorbic acid?

5. Stoichiometry: Standard Solutions

A sample of iron II ion, Fe^{2+}, is standardized with an acid solution of dichromate ion, $Cr_2O_7^{2-}$.

(a) Write the balanced equation for the titration reaction, shown in an unbalanced 'skeleton' form.

$$Fe^{2+} + Cr_2O_7^{2-} \Rightarrow Fe^{3+} + Cr^{3+}$$

(b) A standard is prepared by dissolving 0.1176 g of potassium dichromate, $K_2Cr_2O_7$ (molar mass 294.1), in 100.0 mL of a water solution containing hydrochloric acid. The standard required 28.6 mL of Fe^{2+} solution to completely reduce the $Cr_2O_7^{2-}$. What is the molarity of the Fe^{2+} solution?

(c) It requires 26.9 mL of the standardized Fe^{2+} solution to neutralize 25.0 mL of an acidified permanganate, MnO_4^-, solution. What is the molarity of the MnO_4^- solution?
The balanced equation for the reaction is:

$$5Fe^{2+} + MnO_4^- + 8H^+ \Rightarrow 5Fe^{3+} + Mn^{2+} + 4H_2O$$

6. Thermodynamics

Methane, CH_4, and chlorine, Cl_2, are reacted to form carbon tetrachloride, CCl_4, and hydrogen chloride, HCl at 25°C. The heat of the reaction, $\Delta H°$, is -152.9 kJ per mole of CH_4 reacted. The equation is:

$$CH_4(g) + 2Cl_2(g) \Rightarrow CCl_4(l) + 4HCl(g)$$

Substance	Heat of Formation, $\Delta H_f°$, kJ mol^{-1}	Absolute Entropy, $S°$, J mol^{-1} K^{-1}
$C_{graphite}(s)$	---.--	5.740
$CH_4(g)$	-74.86	186.2
$CCl_4(l)$	?	310.0
$Cl_2(g)$	---.--	223.0
HCl(g)	-92.31	186.8

(a) What is the standard heat of formation, $\Delta H_f°$, for carbon tetrachloride at 25°C?

(b) Calculate the standard entropy change, $\Delta S°$, for the reaction at 25°C.

(c) Theoretically carbon tetrachloride can be formed by the reaction:
$$C_{graphite}(s) + 2Cl_2(g) \Rightarrow CCl_4(l)$$
Calculate the standard free energy of formation, $\Delta G_f°$, for carbon tetrachloride.

(d) Calculate the value of the equilibrium constant, K, for the reaction in Part (c) at 25°C.

Sample Answers and Explanations

1. Electrochemistry: Electrolytic Cells
   (a) Answer: $Pb^{2+} + 2H_2O \Rightarrow PbO_2 + 4H^+ + 2e^-$
   The oxidation reaction occurs at the anode.

   (b) Answer: 2.25 grams
   $$\text{Faradays} = \frac{1.75 \text{ coul}}{1 \text{ sec}} \times 1200 \text{ sec} \times \frac{1 \text{ mol e}^-}{96500 \text{ coul}}$$
   Faradays = 0.0218 mol e⁻

   Cathode (Reduction): $Pb^{2+} + 2e^- \Rightarrow Pb$
   $$g_{Pb} = 0.0218 \text{ mol e}^- \times \frac{1 \text{ mol Pb}}{2 \text{ mole e}^-} \times \frac{207 \text{ g Pb}}{1 \text{ mol Pb}} = 2.25 \text{ g}$$

   (c) Answer: 8.29 grams
   $$\text{Faradays} = 49.5 \text{ g Pb} \times \frac{1 \text{ mol Pb}}{207 \text{ g Pb}} \times \frac{2 \text{ mol e}^-}{1 \text{ mol e}^-}$$
   Faradays = 0.478 mol e⁻

   Cathode (Reduction): $Cr^{3+} + 3e^- \Rightarrow Cr$
   $$g_{Cr} = 0.478 \text{ mol e}^- \times \frac{1 \text{ mol Cr}}{3 \text{ mole e}^-} \times \frac{52.0 \text{ g Cr}}{1 \text{ mol Cr}} = 8.29 \text{ g}$$

2. Electrochemistry: Voltaic Cells
   (a) Answer: $Pd^{2+} + 2e^- \Rightarrow Pd$
   Cathode (Reduction): $Pd^{2+} + 2e^- \Rightarrow Pd$
   Anode (Oxidation): $Cu \Rightarrow Cu^{2+} + 2e^-$

   (b) Answer: $\mathcal{E}° = 0.951$ volts

			$\mathcal{E}°$
Anode (Oxidation):	$Cu \Rightarrow Cu^{2+} + 2e^-$		-0.342 volts
Cathode (Reduction):	$Pd^{2+} + 2e^- \Rightarrow Pd$		x
	$Cu + Pd^{2+} \Rightarrow Pd + Cu^{2+}$		+0.609

   x - 0.342 volts = 0.609 volts $\mathcal{E}° = x = 0.951$ volts

   (c) Answer: $K = 4 \times 10^{20}$

   $\mathcal{E}° = \frac{0.059}{n} \log K$ (At equilibrium the $\mathcal{E}$ in the Nernst equation is 0)

   $0.609 \text{ volts} = \frac{0.0591}{2 \text{ mol}} \log K$

   $K = 4 \times 10^{20}$

   (d) Answer: $\mathcal{E} = 0.557$ volts

   $$\mathcal{E} = \mathcal{E}° - \frac{0.059}{n} \log \frac{[Cu^{2+}]}{[Pd^{2+}]} = 0.609 - \frac{0.059}{2} \log \frac{[3.0]}{[0.0500]}$$

   $\mathcal{E} = 0.557$ volts

158 Free Response Section

3. **Kinetics**

    (a) **Answer: Rate = k[NO]^2[O$_2$]**
    Rate = k[NO]^m[O$_2$]^n

    $\dfrac{\text{Run 2}}{\text{Run 1}} = \dfrac{9.28 \times 10^{-8}}{4.64 \times 10^{-8}} = \dfrac{[\cancel{0.20}]^m}{[\cancel{0.20}]} \times \dfrac{[0.40]^n}{[0.20]}$

    2 = [2]^n n = 1

    $\dfrac{\text{Run 3}}{\text{Run 2}} = \dfrac{3.71 \times 10^{-7}}{9.28 \times 10^{-8}} = \dfrac{[0.40]^m}{[0.20]} \times \dfrac{[\cancel{0.40}]^n}{[\cancel{0.40}]}$

    4 = [2]^m m = 2

    (b) **Answer: k = 5.8 × 10^{-6} M^{-2} sec^{-1}**
    From Run 4 (or any convenient run):
    Rate = 5.80 × 10^{-9} mol L^{-1} sec^{-1} = k[0.10 M]^2[0.10 M]
    k = 5.80 × 10^{-6} M^{-2} sec^{-1}

    (c) **Answer: Rate = K$_1$k$_2$[NO]^2[O$_2$] = k[NO]^2[O$_2$]**
    Rate = k$_2$[NO$_3$][NO]
    K$_1$ = $\dfrac{[NO_3]}{[NO][O_2]}$ [NO$_3$] = K$_1$[NO][O$_2$]
    Rate = K$_1$k$_2$[NO]^2[O$_2$]

    (d) **Answer: 67 kcal = 280 kJ**
    $\ln \dfrac{k_2}{k_1} = \dfrac{-E_a}{R} \left[\dfrac{1}{T_2} - \dfrac{1}{T_1} \right]$
    $\ln \dfrac{5}{1} = -\dfrac{E_a}{1.987 \text{ cal mol}^{-1} \text{ K}^{-1}} \left[\dfrac{1}{1500} - \dfrac{1}{1400} \right]$
    E$_a$ = 67,000 cal = 67 kcal = 280 kJ

4. **Stoichiometry: Empirical Formulas and Colligative Properties**

    (a) **Answer: C$_3$H$_4$O$_3$**

    n$_C$ = n$_{CO_2}$ = 1.500 g ÷ 44.0 g mol^{-1} = 0.0341 mol
    n$_H$ = 2 × n$_{H_2O}$ = (2 × 0.409 g) ÷ 18.0 g mol^{-1}
    n$_H$ = 0.0454 mol

	wt	mol	ratio	empirical
Carbon, C	0.409 g	0.0341 mol	1.00	3
Hydrogen, H	0.045	0.0454	1.33	4
Oxygen, O	0.546	0.0341	1.00	3
Total	1.000 g			

    Empirical formula is C$_3$H$_4$O$_3$

(b) Answer: 175 g mol^{-1}

$$\frac{\text{wt ascorbic}}{1.0 \text{ kg H}_2\text{O}} = \frac{1.00 \text{ g ascorbic}}{10.0 \text{ g H}_2\text{O}} \times \frac{1000 \text{ g H}_2\text{O}}{1 \text{ kg H}_2\text{O}} = 100. \text{ g kg}^{-1}$$

$$m = \frac{\Delta t}{K_f} = \frac{-1.06 \text{ °C}}{-1.86 \text{ °C m}^{-1}} = 0.570 \text{ mol kg}^{-1}$$

Molar mass = 100. g kg^{-1} ÷ 0.570 mol kg^{-1} = 175 g mol^{-1}

(c) Answer: $C_6H_8O_6$

The empirical mass of $C_3H_4O_3$ is 88 g mol^{-1}
# of empirical formulas = molar mass ÷ empirical mass
# of empirical formulas = 176 g mol^{-1} ÷ 88 g mol^{-1} = 2.0
Molecular formula = 2 × $C_3H_4O_3$ = $C_6H_8O_6$

5. Stoichiometry: Standard Solutions

(a) Answer: $6Fe^{2+} + Cr_2O_7^{2-} + 14H^+ \Rightarrow 6Fe^{3+} + 2Cr^{3+} + 7H_2O$

Step	PROCEDURE	
1	HALF–REACTIONS:	$Fe^{2+} \Rightarrow Fe^{3+}$ $Cr_2O_7^{2-} \Rightarrow Cr^{3+}$
2	BALANCE ATOMS, except O and H:	$Fe^{2+} \Rightarrow Fe^{3+}$ $Cr_2O_7^{2-} \Rightarrow 2Cr^{3+}$
3	BALANCE OXYGEN with H_2O:	$Fe^{2+} \Rightarrow Fe^{3+}$ $Cr_2O_7^{2-} \Rightarrow 2Cr^{3+} + 7H_2O$
4	BALANCE HYDROGEN with H^+:	$Fe^{2+} \Rightarrow Fe^{3+}$ $Cr_2O_7^{2-} + 14H^+ \Rightarrow 2Cr^{3+} + 7H_2$
5	BALANCE THE CHARGE with electrons:	$Fe^{2+} \Rightarrow Fe^{3+} + e^-$ $Cr_2O_7^{2-} + 14H^+ + 6e^- \Rightarrow 2Cr^{3+} + 7H_2O$
6	BALANCE ELECTRONS:	$6Fe^{2+} \Rightarrow 6Fe^{3+} + 6e^-$ $Cr_2O_7^{2-} + 14H^+ + 6e^- \Rightarrow 2Cr^{3+} + 7H_2O$
7	COMBINE AND SIMPLIFY: $6Fe^{2+} + Cr_2O_7^{2-} + 14H^+ \Rightarrow 6Fe^{3+} + 2Cr^{3+} + 7H_2O$	

(b) Answer: $[Fe^{2+}]$ = 0.0839 mol L^{-1}

$n_{\text{dichromate}}$ = 0.1176 g ÷ 294.1 g mol-1 = 0.0003999 mol
n_{Fe} = 6 $n_{\text{dichromate}}$ = 0.002399 mol
$[Fe^{2+}]$ = $\frac{0.002399 \text{ mol}}{0.0286 \text{ L}}$ = 0.0839 mol L^{-1}

(c) Answer: $\underline{M}_{MnO_4} = 0.0181$ mol L^{-1}

$$n_{permanganate} = \tfrac{1}{5} n_{Fe}$$
$$\underline{M}_{MnO_4} \times V_{MnO_4} = \tfrac{1}{5} \times \underline{M}_{Fe} \times V_{Fe}$$
$$\underline{M}_{MnO_4} = \tfrac{1}{5} \times \frac{(0.0839 \text{ mol L}^{-1} \times 26.9 \text{ mL})}{25.0 \text{ mL}}$$
$$\underline{M}_{MnO_4} = 0.0181 \text{ mol L}^{-1}$$

6. <u>Thermodynamics</u>

   (a) **Answer:** $\Delta H_f° = +141.5$ kJ mol^{-1}

   $\Delta H° = \Sigma \Delta H_f°$ (Products) $- \Sigma \Delta H_f°$ (Reactants)
   -152.9 kJ $= (\Delta H_f° + (4 \times -92.31$ kJ$)) - (-74.86 + 0)$
   $\Delta H_f° = +141.5$ kJ mol^{-1}

   (b) **Answer:** $\Delta S° = +648.0$ J mol^{-1} K^{-1}

   $\Delta S° = \Sigma S_f°$ (Products) $- \Sigma S_f°$ (Reactants)
   $\Delta S° = (310.0 + (4 \times 186.8)) - (186.2 + (2 \times 223.0))$
   $\Delta S° = +425.0$ J mol^{-1} K^{-1}

   (c) **Answer:** $\Delta G_f° = -99.2$ kJ mol^{-1}

   $$\Delta G° = \Delta H° - \frac{T \Delta S°}{1000}$$

   $\Delta H° = \Delta H_f°_{CCl_4} = +141.5$ kJ mol^{-1} (from part (a))

   $\Delta S° = \Sigma S_f°$ (Products) $- \Sigma S_f°$ (Reactants)
   $\Delta S° = ((310.0) - (5.74 + (2 \times 223.0)))$
   $\Delta S° = -141.7$ J mol^{-1} K^{-1}

   $$\Delta G° = +141.5 - \frac{((298 \text{ K}) \times (-141.7 \text{ J mol}^{-1} \text{ K}^{-1}))}{1000}$$
   $\Delta G° = +184$ kJ mol^{-1}

   (d) **Answer: K = 6 × 10^{-33}**

   $\Delta G° = -RT \ln K$
   $\ln K = -\dfrac{(+184{,}000 \text{ J mol}^{-1})}{(8.314 \text{ J mol}^{-1} \text{ K}^{-1} \times 298 \text{ K})} = -74.3$
   $K = e^{-74.3}$ ('INVERSE' ln -74.3) $= 6 \times 10^{-33}$

Chapter 3

Part C: Equations

This part requires that you write 5 reactions (chosen from 8) as <u>in net ionic form</u>. There will always be a reaction. The products will be different from the reactants in every case.

Each reaction is worth 3 points. One (1) point is given for writing the reactant formulas correctly. Up to two (2) points are given as (full or partial) credit for correctly predicting the products. Remember; to receive this credit the equation <u>must</u> be in net ionic form.

Do only 5. If you do more, the <u>first 5</u> will be scored. This part is worth 15 points, and should be given only 10–15 minutes.

The equations do not have to be balanced, and the states are not required. Balancing the equations and putting in the states will sometimes help predict the products, and there is no rule against it.

1. **General Considerations**

    a. Common polyatomic anions and cations.

$C_2H_3O_2^-$	acetate	$HCOO^-$	formate
$C_2O_4^{2-}$	oxalate	MnO_4^-	permanganate
ClO^-	hypochlorite	NH_4^+	ammonium
ClO_2^-	chlorite	NO_2^-	nitrite
ClO_3^-	chlorate	NO_3^-	nitrate
ClO_4^-	perchlorate	OH^-	hydroxide
CN^-	cyanide	PO_3^{3-}	phosphite
HCO_3^-	bicarbonate	PO_4^{3-}	phosphate
CO_3^{2-}	carbonate	HSO_3^-	bisulfite
$Cr_2O_7^{2-}$	dichromate	SO_3^{2-}	sulfite
CrO_4^{2-}	chromate	SO_4^{2-}	sulfate

    b. An isolated element which is not an ion has no charge.
    Na° Mg° Cu° Cl_2°

    c. There are 7 diatomic elements. (Diatomic R. **BrINClHOF**)
    Bromine, Br_2; iodine, I_2; nitrogen, N_2; chlorine, Cl_2; hydrogen, H_2; oxygen, O_2; and fluorine, F_2.

2. Solubility Rules for Salts in Water Solutions

The solubility rules 1–6 must be <u>applied in order</u>. The earlier numbered rule <u>takes precedence</u> if more than one rule applies to a salt.

1. **SOLUBLE:** All ammonium, NH_4, and Group IA, Li, Na, K, Rb, Cs, salts.

2. **SOLUBLE:** All nitrates, NO_3^-, acetates, $C_2H_3O_2^-$, and perchlorates, ClO_4^-.

<u>When Rules 1 and 2 do not apply</u>:
3. **INSOLUBLE:** All silver, Ag^+, lead, Pb^{2+}, and mercury(I), Hg_2^{2+}, salts.

<u>When Rule 3 does not apply:</u>
4. **SOLUBLE.** All chlorides, Cl^-, bromides, Br^-, and iodides, I^-.

<u>When Rules 1,2 and 4 do not apply</u>:
5. **INSOLUBLE:** All carbonates, CO_3^{2-}, chromates, CrO_4^{2-}, phosphates, PO_4^{3-} sulfides, S^{2-}, oxides, O^{2-}, and hydroxides, OH^-.

   (Group IIA chromates, except $BaCrO_4$, are <u>soluble</u>.)
   (Group IIA hydroxides, except $Mg(OH)_2$, are <u>soluble</u>.)

<u>When Rules 3 and 5 do not apply:</u>
6. **SOLUBLE:** All sulfates, SO_4^{2-}, except barium sulfate, $BaSO_4$, calcium sulfate, $CaSO_4$ and strontium sulfate, $SrSO_4$.

With few exceptions, the solubility rules permit assuming:

1. A salt with an anion of 1^- charge is SOLUBLE.
   These salts will completely dissociate into ions.

2. Salts with anions of 2^- or 3^- charge are INSOLUBLE.
   Except for certain sulfates, these salts will precipitate.

<u>Examples:</u>
1. Aqueous solutions of silver nitrate and sodium iodide are mixed.
   Silver iodide, AgI, is insoluble (rule 3 ahead of rule 4)
   Sodium nitrate, $NaNO_3$, is soluble (rule 1).
   Complete equation: $Ag^+ + \cancel{NO_3^-} + \cancel{Na^+} + I^- \Rightarrow AgI + \cancel{NO_3^-} + \cancel{Na^+}$
   Net ionic equation: $Ag^+ + I^- \Rightarrow AgI$

2. Solid aluminum chloride is added to an aqueous solution of potassium chromate.
   Aluminum chloride is soluble (Rule 4).
   Potassium chloride is soluble (Rule 1).
   Aluminum chromate is insoluble (Rule 5).
   Complete equation: $AlCl_3 + \cancel{K^+} + CrO_4^{2-} \Rightarrow Al_2(CrO_4)_3 + \cancel{K^+} + Cl^-$
   Net ionic equation: $AlCl_3 + CrO_4^{2-} \Rightarrow Al_2(CrO_4)_3 + Cl^-$

3. Reactions Involving No Changes in Oxidation States

<u>Displacement reactions.</u>

The products can be predicted by exchanging the positive ions of the two reactants.

<u>Example:</u>
Water solutions of hydrochloric acid and sodium hydroxide are mixed.
$$HCl + NaOH \Rightarrow NaCl + HOH$$

HCl is a strong acid, NaOH is a strong base, and NaCl is a soluble salt; these should be written as ions.
$$H^+ + Cl^- + Na^+ + OH^- \Rightarrow Na^+ + Cl^- + HOH$$

Substances that do not change ('spectator ions') are not represented in a net ionic equation.
$$H^+ + \cancel{Cl^-} + \cancel{Na^+} + OH^- \Rightarrow \cancel{Na^+} + \cancel{Cl^-} + HOH$$
$$H^+ + OH^- = HOH$$

<u>Combination reactions</u> (acid and base anhydrides)

<u>Examples:</u>
1. Metal oxide + water ⇒ a base
$$CaO(s) + HOH \Rightarrow Ca(OH)_2 \text{(soluble hydroxide)} \Rightarrow Ca^{2+} + 2OH^-$$

2. Nonmetal oxide + water ⇒ an acid
$$SO_2 + HOH \Rightarrow H_2SO_3 \text{ (not a strong acid)}$$

3. Metal oxide + nonmetal oxide = salt
$$CaO + SO_2 \Rightarrow CaSO_3$$

<u>Decomposition reactions</u> (the reverse of combination reactions)

<u>Examples:</u>
1. Base ⇒ metal oxide + water
$$Ca(OH)_2 \Rightarrow CaO + HOH$$
$$Mg(OH)_2 \Rightarrow MgO + H_2O$$

2. Acid containing oxygen ⇒ nonmetal oxide + water
$$H_2CO_3 \Rightarrow CO_2 + HOH$$

3. Salt containing oxygen ⇒ metal oxide + nonmetal oxide
$$CaCO_3 \Rightarrow CaO(s) + CO_2$$

Hydrolysis Reactions

Hydrolysis occurs when water reacts with a compound. The water is written as HOH. Combining the 'H+' from the water with negative ion from the other reactant usually gives the formula for one of the products.

Example:
1. Sodium acetate is dissolved in water.

    $NaC_2H_3O_2 + HOH \Rightarrow HC_2H_3O_2 + NaOH$

2. Sodium salts are soluble in water.
   $HC_2H_3O_2$ is not a strong acid and is not ionized.

    $\cancel{Na^+} + C_2H_3O_2^- + HOH \Rightarrow HC_2H_3O_2 + \cancel{Na^+} + OH^-$

3. Omit the 'spectator ions' to get the net ionic equation.

    $C_2H_3O_2^- + HOH \Rightarrow HC_2H_3O_2 + OH^-$

Reactions of coordination compounds and ions

The ligands (Lewis acids) are bonded to a central atom that is usually a transition metal ion. The most frequently occurring ligands are ammonia, NH_3, and hydroxide ion, OH^-.

Ammonia	'Excess' hydroxide
$Ag(NH_3)_2^{1+}$	$Al(OH)_4^{1-}$
$Cu(NH_3)_4^{2+}$	$Zn(OH)_4^{2-}$
$Ni(NH_3)_6^{2+}$	$Cr(OH)_6^{3-}$

The number of ligands attached to a central metal ion is sometimes twice the oxidation number of the central metal.

$Ag(NH_3)_2^+, Zn(OH)_4^{2-}, Fe(CN)_6^{3-}$

The breakup of complex ions is frequently achieved by adding an acid. The products are the metal ion and the species formed when hydrogen ions from the acid react with the ligand (a Lewis base).

Example:
Tetraammine copper II ions are reacted with nitric acid.

$Cu(NH_3)_4^{2+} + H^+ \Rightarrow Cu^{2+} + NH_4^+$

Nonaqueous definitions of acids and bases.
a. Brönsted reactions involve the transfer of a proton.
b. Lewis reactions involve the formation of a coordinate covalent bond.

Example of a Lewis reaction:
The gases boron trifluoride and ammonia are mixed.

$BF_3 + NH_3 \Rightarrow BF_3NH_3$

Part C: Equations

4. Redox Reactions (Change in Oxidation State)

IMPORTANT OXIDIZERS:

OXIDIZER:	Formed in the reaction:
MnO_4^- in acid	Mn^{2+}
MnO_2 in acid	Mn^{2+}
MnO_4^- in neutral <u>or</u> basic solution	MnO_2
$Cr_2O_7^{2-}$ in acid	Cr^{3+}
HNO_3, concentrated	NO_2
HNO_3, dilute	NO
H_2SO_4, hot, conc.	SO_2
Metal-ic ions	metal-ous ions
Free halogens	halide ions
Na_2O_2	$NaOH$
$HClO_4$	Cl^-

IMPORTANT REDUCERS:

REDUCERS:	Formed in the reaction:
Halide ions	Free halogen
Free metals	Metal ions
Sulfite ions (SO_2(aq))	Sulfate ions
Nitrite ions	Nitrate ions
Free halogens, dil. basic sol'n.	Hypohalite ions
Free halogens, conc. basic sol'n.	Halite ions
Metal-ous ions	Metal-ic ions

<u>Redox:</u> <u>Reactions between an oxidizer and a reducer.</u>

Redox reactions are often recognized by:

....familiarization with important reducers and oxidizers.

....the clue that there is 'added acid' or the solution is 'acidified'.

....the use of the supplied reduction potential reference.

<u>Examples:</u>

1. Manganese dioxide is added to concentrated hydrochloric acid and heated.

    $MnO_2 + H^+ + Cl^- \Rightarrow Mn^{2+} + Cl_2 + H_2O$

2. A solution of iron(II) nitrate is added to an acidified solution of potassium permanganate.

    $Fe^{2+} + H^+ + MnO_4^- \Rightarrow Fe^{3+} + Mn^{2+} + H_2O$

3. Magnesium metal is added to dilute nitric acid. One of the products contains nitrogen with an oxidation number of -3.

    $Mg + H^+ + NO_3^- \Rightarrow Mg^{2+} + NH_3 + H_2O$

Redox: Combination reactions.

An oxidizer will react with a reducer of the same element to produce the element at an intermediate oxidation state.

Examples:
1. Solutions of potassium iodide, potassium iodate, and dilute sulfuric acid are mixed.

   $I^- + IO_3^- + H^+ \Rightarrow I_2 + H_2O$

2. A piece of iron is added to a solution of iron (III) sulfate.

   $Fe + Fe^{3+} \Rightarrow Fe^{2+}$

Redox: Replacement reactions.

A more reactive element (often in the free state) can displace a less reactive element with similar properties from a compound.

Examples:
1. Zinc metal reacts with tin (II) sulfate.

   $Zn + Sn^{2+} \Rightarrow Zn^{2+} + Sn$

2. Free chlorine reacts with sodium bromide.

   $Cl_2 + Br^- \Rightarrow Cl^- + Br_2$

3. Solid barium peroxide is added to cold sulfuric acid.

   $BaO_2 + H^+ + SO_4^{2-} \Rightarrow BaSO_4 + H_2O_2$

Redox: Decomposition reactions

Examples:
1. A solution of hydrogen peroxide is catalytically decomposed.

   $H_2O_2 \Rightarrow H_2O + O_2$

2. Chlorates decompose in the presence of heat.

   $KClO_3 \Rightarrow KCl + O_2$

3. Electrolysis decomposes compounds into their elements.

   $H_2O \Rightarrow H_2 + O_2$

Part C: Equations

5. Practice Equation Writing

Write the formula equations for each reaction in net ionic format. Do not balance. There is a reaction in each case.

Select the reaction type that will lead to the correct prediction of the products.

1. Sodium sulfite crystals are added to water.

2. Solid potassium dichromate is added to an aqueous solution of lead(II) nitrate.

3. Calcium carbonate chips are added to excess nitric acid.

4. Hydrogen sulfide gas is bubbled through a solution of cadmium nitrate.

5. Solutions of acetic acid and sodium carbonate are mixed.

6. Hydrogen chloride gas is bubbled through water.

7. Solutions of carbon dioxide gas and ammonia gas are mixed.

8. Solid lithium oxide is added to water.

9. Sulfur trioxide gas is bubbled through a solution of sodium hydroxide.

10. A mixture of the solid calcium oxide and solid tetraphosphorous decaoxide is heated.

11. Solid magnesium carbonate is heated in a crucible.

12. Solid phosphorus trichloride is added to water.

13. Solid sodium cyanide is added to water.

14. Solid zinc nitrate is treated with excess sodium hydroxide solution.

15. An aqueous solution of diamminesilver chloride is treated with dilute nitric acid.

16. A suspension of copper(II) hydroxide in water is treated with an excess of ammonia.

17. Gaseous boron hydride gas is mixed with ammonia gas.

18. Boron trifluoride is added to gaseous trimethylamine.

19. Gaseous silane (SiH_4) is burned in excess oxygen.

20. Magnesium ribbon is burned in pure nitrogen gas.

21. An aqueous hydrogen peroxide solution is added to an acidified solution of potassium iodide.

22. Dilute potassium permanganate solution is added to an oxalic acid solution which was acidified with a few drops of sulfuric acid.

23. A solution containing tin(II) ion is added to acidified potassium dichromate.

24. A segment of copper wire is added to dilute nitric acid.

25. Sodium dichromate is added to an acidified solution of sodium iodide.

26. Sodium hydride crystals are added to water.

27. Powdered iron is added to a solution of iron(III) sulfate.

28. Chlorine gas is bubbled through a solution of potassium bromide.

29. Solid sodium is added to water.

30. A dilute solution of sulfuric acid is electrolyzed between platinum electrodes.

Part C: Equations

Sample Answers and Explanations

1. **EQUATION:** $Na_2SO_3 \Rightarrow Na^+ + HSO_3^- + OH^-$

   REACTION TYPE: Solubility and hydrolysis

   Notes: Sodium salts are soluble in water (Rule 1).
   Bases react with water for form OH^-.

2. **EQUATION:** $K^+ + Cr_2O_7^{2-} + Pb^{2+} + \cancel{NO_3^-} \Rightarrow PbCr_2O_7 + \cancel{K^+} + \cancel{NO_3^-}$
   NET IONIC: $Cr_2O_7^{2-} + Pb^{2+} \Rightarrow PbCr_2O_7$

   REACTION TYPE: Displacement.

   Notes: Potassium (Rule 1) and nitrate (Rule 2) salts dissolve completely. K^+ and NO_3^- ions are 'spectators'.

3. **EQUATION:** $CaCO_3(s) + H^+ + \cancel{NO_3^-} \Rightarrow Ca^{2+} + \cancel{NO_3^-} + H_2CO_3$
   NET IONIC: $CaCO_3(s) + H^+ \Rightarrow Ca^{2+} + CO_2 + H_2O$

   REACTION TYPE: Displacement.

   Notes: HNO_3 is a strong acid and completely ionizes.
   Carbonic acid, H_2CO_3, decomposes in water to give CO_2 gas bubbles.

4. **EQUATION:** $Cd^{2+} + \cancel{NO_3^-} + H_2S \Rightarrow CdS + H^+ + \cancel{NO_3^-}$
   NET IONIC: $Cd^{2+} + H_2S \Rightarrow CdS + H^+$

   REACTION TYPE: Displacement.

   Notes: Nitrate salts are soluble (Rule 2).
   Many sulfides are insoluble (Rule 5).

5. **EQUATION:** $HC_2H_3O_2 + \cancel{Na^+} + HCO_3^- \Rightarrow C_2H_3O_2^- + \cancel{Na^+} + H_2CO_3$
   NET IONIC: $HC_2H_3O_2 + HCO_3^- \Rightarrow C_2H_3O_2^- + CO_2 + H_2O$

   REACTION TYPE: Displacement

   Notes: Acid-base (Brönsted proton transfer).
   H_2CO_3 decomposes to form CO_2 and H_2O.

6. **EQUATION:** $HCl + HOH \Rightarrow H_3O^+ + Cl^-$

   REACTION TYPE: Displacement

   Notes: Acid-base (Brönsted proton transfer).

7. **EQUATION:**
   1. $CO_2 + H_2O \Leftrightarrow H_2CO_3$
   2. $NH_3(aq) + H_2CO_3(aq) \Rightarrow NH_4^+ + HCO_3^-$

   NET IONIC: $CO_2 + H_2O + NH_3(aq) \Rightarrow NH_4^+ + HCO_3^-$

   REACTION TYPE: Displacement

   Notes: Acid-base (Brönsted proton transfer). H_2CO_3, carbonic acid, is $CO_2 + H_2O$.

8. **EQUATION:** $Li_2O + HOH \Rightarrow LiOH(aq) \Rightarrow Li^+ + OH^-$
   NET IONIC: $Li_2O + HOH \Rightarrow Li^+ + OH^-$

   REACTION TYPE: Combination.

   Notes: LiOH is a strong base and completely dissociates. Li_2O is a basic anhydride.

9. **EQUATION:**
   1. $SO_3(g) + H_2O \Rightarrow H_2SO_4 \Rightarrow H^+ + HSO_4^-$
   2. $H^+ + \cancel{Na^+} + OH^- \Rightarrow HOH + \cancel{Na^+}$

   NET IONIC: $SO_3 + OH^- \Rightarrow HSO_4^{2-}$

   REACTION TYPE: Combination
   1. Nonmetal oxide in water gives an acid.
   2. Acid-base (Brönsted proton transfer).

   Notes: H_2SO_4 is a strong acid. Sodium hydroxide is a strong base.

10. **EQUATION:** $CaO(s) + P_4O_{10}(s) \Rightarrow Ca_3(PO_4)_2(s)$

    REACTION TYPE: Combination.

    Notes: Metal oxide + nonmetal oxide = salt. There is <u>no</u> change in oxidation number.

11. **EQUATION:** $MgCO_3 \Rightarrow MgO + CO_2$

    REACTION TYPE: Decomposition.

    Notes: Heating a carbonate yields a metal oxide and carbon dioxide.

12. **EQUATION:** $PCl_3 + HOH \Rightarrow H^+ + Cl^- + P(OH)_3$
    OR $PCl_3 + HOH \Rightarrow H^+ + Cl^- + H_3PO_3$

    REACTION TYPE: Hydrolysis or displacement.

    Notes: HCl is a strong acid and completely ionizes.

Part C: Equations   171

13. **EQUATION:** NaCN + HOH $\Rightarrow$ HCN + Na$^+$ + OH$^-$

    REACTION TYPE: Hydrolysis of a base

    Notes: HCN is not a strong acid.

14. **EQUATION:** Zn(NO$_3^-$)$_2$ + ~~Na$^+$~~ + OH$^-$ $\Rightarrow$ Zn(OH)$_4^{2-}$ + ~~Na$^+$~~ + NO$_3^-$
    NET IONIC: Zn(NO$_3^-$)$_2$ + OH$^-$ $\Rightarrow$ Zn(OH)$_4^{2-}$ + NO$_3^-$

    REACTION TYPE: Reactions of coordination compounds and ions.

    Notes: Key words are 'excess sodium hydroxide'.
    Sodium and nitrate salts are soluble.

15. **EQUATION:** Ag(NH$_3$)$_2^+$ + Cl$^-$ + H$^+$ + ~~NO$_3^-$~~ $\Rightarrow$ AgCl + NH$_4^+$ + ~~NO$_3^-$~~
    NET IONIC: Ag(NH$_3$)$_2^+$ + Cl$^-$ + H$^+$ $\Rightarrow$ AgCl + NH$_4^+$

    REACTION TYPE: Reactions of complex ions.

    Notes: Breakup of complex ions is frequently achieved by adding an acid which reacts with the basic ligand.
    Many silver salts are insoluble (Rule 3).

16. **EQUATION:** Cu(OH)$_2$(s) + NH$_3$(aq) $\Rightarrow$ Cu(NH$_3$)$_4^{2+}$ + OH$^-$

    REACTION TYPE: Reactions of coordination compounds.

    Notes: Key words are 'excess ammonia'.

17. **EQUATION:** BH$_3$ + NH$_3$ $\Rightarrow$ BH$_3$NH$_3$

    REACTION TYPE: Lewis acids and bases.

    Notes: BH$_3$ has 6 outer electrons. NH$_3$ has 8 including a 'lone pair'.
    A coordinate covalent bond is formed.

18. **EQUATION:** BF$_3$ + (CH$_3$)$_3$N $\Rightarrow$ F$_3$BN(CH$_3$)$_3$

    REACTION TYPE: Lewis acids and bases.

    Notes: BF$_3$ has 6 outer electrons. (CH$_3$)$_3$N has 8 including a 'lone pair'.
    A coordinate covalent bond is formed.

19. **EQUATION:** $SiH_4 + O_2 \Rightarrow SiO_2 + H_2O$

    REACTION TYPE: Reaction between reducer and oxidizer.

    Notes: Combustion forms the oxides of elements.
    In excess oxygen the highest oxidation state oxide forms.

20. **EQUATION:** $Mg(s) + N_2(g) \Rightarrow Mg_3N_2(s)$

    REACTION TYPE: Reaction between reducer and oxidizer. Redox combination.

    Notes: Metals react with nonmetals to form salts.

21. **EQUATION:** $H_2O_2 + H^+ + \cancel{K^+} + I^- \Rightarrow H_2O + \cancel{K^+} + I_2$
    NET IONIC: $H_2O_2 + H^+ + I^- \Rightarrow H_2O + I_2$

    REACTION TYPE: Reaction between oxidizer and reducer.

    Notes: Potassium salts are soluble (Rule 1).
    Peroxide in acid is an important oxidizer.
    Halide ions are important reducers.

22. **EQUATION:** $MnO_4^- + H^+ + H_2C_2O_4 \Rightarrow Mn^{2+} + CO_2 + H_2O$

    REACTION TYPE: Reactions between oxidizer and reducer.

    Notes: Permanganate in acid is an important oxidizer.
    Oxalic acid is not a strong acid.
    H_2CO_3 decomposes to form CO_2 and H_2O.

23. **EQUATION:** $Sn^{2+} + H^+ + Cr_2O_7^{2-} \Rightarrow Sn^{4+} + Cr^{3+} + H_2O$

    REACTION TYPE: Reaction between reducer and oxidizer.

    Notes: Dichromate in acid is an important oxidizer.

24. **EQUATION:** $Cu + H^+ + NO_3^- \Rightarrow Cu^{2+} + NO + H_2O$

    REACTION TYPE: Reaction between reducer and oxidizer.

    Notes: Nitric acid both a strong acid and an important oxidizer.

Part C: Equations 173

25. EQUATION: $\cancel{Na^+} + Cr_2O_7^{2-} + H^+ + I^- \Rightarrow \cancel{Na^+} + Cr^{3+} + I_2 + H_2O$
    NET IONIC: $Cr_2O_7^{2-} + H^+ + I^- \Rightarrow Cr^{3+} + I_2 + H_2O$

    REACTION TYPE: Reaction between oxidizer and a reducer.

    Notes: Dichromate <u>in acid</u> is an important oxidizer.
    Sodium salts are soluble in water (Rule 1).

26. EQUATION: $NaH(s) + HOH \Rightarrow Na^+ + OH^- + H_2(g)$

    REACTION TYPE: Redox combination
    Metal hydride + HOH $\Rightarrow$ base + hydrogen.

    Notes: NaOH completely dissociates.

27. EQUATION: $Fe + Fe^{3+} \Rightarrow Fe^{2+}$

    REACTION TYPE: Redox combination.

    Notes: Fe^{2+} is an intermediate oxidation state.

28. EQUATION: $Cl_2 + \cancel{K^+} + Br^- \Rightarrow \cancel{K^+} + Cl^- + Br_2$
    NET IONIC: $Cl_2 + Br^- \Rightarrow Cl^- + Br_2$

    REACTION TYPE: Redox replacement.

    Notes: A more reactive element displaces a less reactive element in a compound.
    Potassium salts completely dissolve (Rule 1).

29. EQUATION: $Na + HOH \Rightarrow Na^+ + OH^- + H_2$

    REACTION TYPE: Redox displacement.

    Notes: A more reactive element displaces a less reactive element in a compound.
    Strong bases completely dissociate.

30. EQUATION: $H_2O \Rightarrow H_2 + O_2$

    REACTION TYPE: Redox decomposition. (Electrolysis)

    Notes: Electrolysis decomposes compounds into their elements. In this case the voltage required to decompose water is less than the voltage required to decompose H_2SO_4.

Chapter 4
Part D: Essays

The essays require an ability to present ideas in a logical order and making use of defined chemical principles to explain observations that are cited in the question. Part D permits the choice of three out of five topics, and is worth 35 percent. If you do more than three questions, only the first three will be scored. About 40 minutes can be spent on this part, or about 10-15 minutes for each question.

In order to conserve time, the answers to the essays should be as concise as possible. For these questions:

1. Think before you start writing.

2. Circle the words that tell you what to do (i.e. compare, contrast, describe, explain).

3. Define the chemical terms used in the question. Most essays will be answered by using defined words properly.

4. Be organized. Start with a thesis statement, then supporting points and finish with a conclusion.

5. Be specific. Use only relevant facts in your answer.

6. Stick to the point of the question. Points can be lost by giving an answer that is too general.

7. Limit yourself to one thought per sentence. The sentences will end up being different lengths and the result will be in a more conversational tone.

8. Do not get hung up so long on a question that time is not available for other questions.

9. Do not change essay answers if the time is running out. When one sentence is changed there is a good chance that other sentences need to be changed also.

Year	Part D				
1976	*Atom* Periodicity	*Bond* Octets	*Thermo* ΔH, ΔS, ΔG	*States* Ideal Gases	*Kinetics* Rate
1977	*Atom* Affinity	*Bond* Major types	*Bond* Coordinates	*Bond* Isomers	*Equilibrium* LeChatelier
1978	*Atom* Bohr Model	*Atom/Bond* Polarity, I.E.	*Bond* Isomers	*Stoich* Colligative	*Equilibrium* Hydrolysis
1979	*Bond* Shapes	*Bond* Dipoles	*Laboratory* Analysis	*Kinetics* Rates	*Equilibrium* Acid-Base
1980	*Atom* Structure	*Laboratory* Solutions	*Kinetics* Pot. Energy	*Equilibrium* LeChatelier	*Thermo* Entropy
1981	*Atom* Spectra	*Stoich* Empiric	*Equilibrium* Hydrolysis	*Thermo* Free Energy	*Elect–chem* Electroysis
1982	*Atom* Periodicity	*Bond* VSEPR	*States* Ideal	*Laboratory* Solutions	*Equilibrium* Titration
1983	*Bond* Coordinates	*Bond* Molec. Orb.	*Equilibrium* Buffers	*Kinetics* Order	*Elect–chem* Voltaic cell
1984	*Atom* Metals	*States* Vap. Press.	*Laboratory* Solutions	*States* Gases	*Equilibrium* Titration
1985	*Atom* Periodicity	*Bond* Dipoles	*Laboratory* Solutions	*Kinetics* Rate Law	*Thermo* ΔH, ΔS, ΔG
1986	*Atom* I.E., ΔH	*Bond* Properties	*Bond* Oxyacids	*Kinetics* Rate Law	*Elect–chem* Reactions
1987	*Atom* Uncertainty	*Atom* Periodicity	*Bond* Dissociation	*Thermo* ΔH, ΔS, ΔG	*Elect–chem* Electrolysis
1988	*Bond* Properties	*States* Phases	*Equilibrium* LeChatelier	*Equilibrium* Titration	*Thermo* Lab ΔH
1989	*Atom* Decay	*Bond* Shapes	*Bond* Melt Properties	*Laboratory* Metal Reactions	*Kinetics* Reaction Rate

The table shows the questions asked in Part D for the past several years. The essays chosen balance the topics in Parts A and B so that all the major areas are tested in a somewhat balanced manner in the three parts.

Sample Essay Questions

1. An empirical rule commonly known as "Hund's Rule" states that "in a given atom, so long as the Pauli Exclusion Principle permits, electrons in the same subshell will occupy orbitals with different values of m and their spins will not pair up."

    (a) What is the Pauli Exclusion Principle?

    (b) Explain how Hund's Rule is applied to the electron structure of many–electron atoms.

    (c) Some atoms are paramagnetic and others are diamagnetic. What is the difference between the electronic structure of elements which are paramagnetic and those that are diamagnetic?

    (d) Hund's Rule has been confirmed by measurements of the magnetic properties of elements. Using selected Period 2 elements as examples, show how the confirmation of Hund's Rule has been accomplished.

2. The electron affinities, in kJ mol^{-1}, are given for five elements in Period 2.

B = -23	C = -123	N = 0	O = -142	F = -322

    (a) Why does the electron affinity of the elements generally increase with atomic number in Period 2.

    (b) Why is the electron affinity of nitrogen zero, while carbon's and oxygen's are substantial?

3. In a P$_4$ molecule, the phosphorus atoms are at the corners of a tetrahedron. Each phosphorus atom is bonded to the three others. All the bonds have the same energy.

    (a) Draw a diagram showing the shape of a P$_4$ molecule.

    (b) What is the bond angle for each P—P—P bond?

    (c) Draw a diagram showing the shape of a PH$_3$ molecule.

    (d) Account for the difference between P—P—P bond angle and the H—P—H bond angle.

4. The strength of bonding of ligands in complex ions decreases in the order:
$CN^- > NO_2^- > NH_3 > H_2O > OH^- > F^- > Cl^- > Br^- > I^-$. Thus cyanide ion, CN^-, is at the head of the list and is said to be a very strong ligand. At the other end iodide ion, I^-, is a very weak ligand.

(a) Draw the Lewis electron–dot structures of the cyanide ion and the nitrite ion.

(b) How many sigma, σ, and how many pi, π, bonds are in:

......a cyanide ion?

......a nitrite ion?

(c) What characteristic of the structure of ligands causes them to form complex ions?

(d) Explain where carbon monoxide, CO, would fit into the series of bonding strengths of ligands.

5. Iodate ion, IO_3^-, can be used to titrate tantalum (I) ion, Tl^+, in a concentrated solution of HCl.

$$IO_3^- + Tl^+ + Cl^- \Rightarrow ICl_2^- + Tl^{3+}$$

(a) Write the correctly balanced:

(1) oxidation half–reaction

(2) reduction half–reaction

(3) net ionic reaction.

(b) Which is the reducing agent and which is the oxidizing agent in the reaction?

(c) One of the reactants can be made 1.50 M, and one of the products can be made 0.10 M. Which reactant and which product should be chosen to maximize the net cell voltage, $\mathcal{E}°$? Explain your reasoning.

(d) A measurement or a calculation of the net cell voltage, $\mathcal{E}°$, can be used to determine two additional properties of the reaction. List the two properties of the reaction that would be determined.

6. Four beakers containing 0.10 M solutions are prepared.
   AgNO₃   Al(NO₃)₃   LiNO₃   Zn(NO₃)₂

   (a) Six-molar (6.0 M) sodium hydroxide, NaOH, is added to each solution. Write balanced ionic equations for the solutions where a precipitation reaction occurs.

   (b) Excess NaOH is added to the beakers where a precipitate has formed. Write balanced ionic equations for the precipitate(s) that undergo further reaction.

7. Consider the system given to be at equilibrium, 25°C and a constant volume of 1.0-Liter.

   $$C_{graphite}(s) + 2H_2(g) \Leftrightarrow CH_4(g) \quad \Delta H° = -75 \text{ kJ}$$

   State whether the molarity, M, of each of the three constituents of the system shows a net increase, a net decrease or remains constant as the system comes back to equilibrium.

   (a) Increase the temperature.

   (b) Remove some hydrogen, H₂.

   (c) Add some graphite, C.

   (d) Decrease the pressure.

   (e) Add some methane, CH₄.

8. The weight percent of NaHCO₃ in a mixture of sodium hydrogen carbonate, NaHCO₃, and sodium chloride, NaCl, was determined by heating the mixture. Upon heating, the mixture lost weight equal to the weight of the carbon dioxide and water released by the NaHCO₃.

   (a) Write a balanced reaction equation.

   (b) If time was not taken to bring the residue to constant weight by heating the sample long enough, would the reported percent sodium bicarbonate be too high or too low? Explain your conclusion.

   (c) Before the sample can be weighed, it must be cooled to about room temperature. Would the reported percent sodium bicarbonate be too high or too low if the cooling process were allowed to occur in the room overnight?
   Explain your conclusion.

9. Phosgene, $COCl_2$, is produced in the gas phase from carbon monoxide, CO, and chlorine, Cl_2.

$$CO(g) + Cl_2(g) \Rightarrow COCl_2(g)$$

The reaction proceeds by the mechanism:

1.	Cl_2	$\Leftrightarrow$ 2Cl	(fast equilibrium)
2.	Cl + CO	$\Leftrightarrow$ COCl	(fast equilibrium)
3.	$COCl + Cl_2$	$\Rightarrow COCl_2 + Cl$	(slow)
4.	2Cl	$\Leftrightarrow Cl_2$	(fast equilibrium)

(a) Write the rate law equation in terms of substances that appear in the overall reaction.

(b) What are the units of the rate law constant, k?

(c) An increase of 10°C will possibly double the rate of a chemical reaction. Explain what occurs on the molecular level that accounts for this rate change.

(d) The value of the rate constant, k, can be graphed at different Kelvin temperatures, T, using data obtained in the laboratory. From this data the activation energy, E_a, can be determined.

1. What should be plotted as the independent variable (x–axis) and what should be plotted as the dependent variable (y–axis)?

2. How would the graph be used to determine the activation energy for the reaction?

10. For the reaction:

$$Fe_2O_3(s) + 3C(s) \Leftrightarrow 2Fe(s) + 3CO(g)$$

$\Delta H°$, $\Delta S°$, and $\Delta G°$ are all positive when the substances are in their standard states at 25°C.

(a) What is the physical significance of the signs of $\Delta H°$, $\Delta S°$, and $\Delta G°$ for this reaction?

(b) Which of the substances would exist in the highest amounts in an equilibrium mixture at 25°C? Explain how you arrived at your answer.

(c) This reaction is used as a step in the recovery of iron from its ore. Use thermodynamic concepts to explain how the yield of iron can be maximized.

11. Deviations from ideal behavior for nitrogen, N_2, hydrogen, H_2, and carbon dioxide, CO_2, are shown in the graph. Ideal gases have a PV/nRT ratio equal to 1.0.

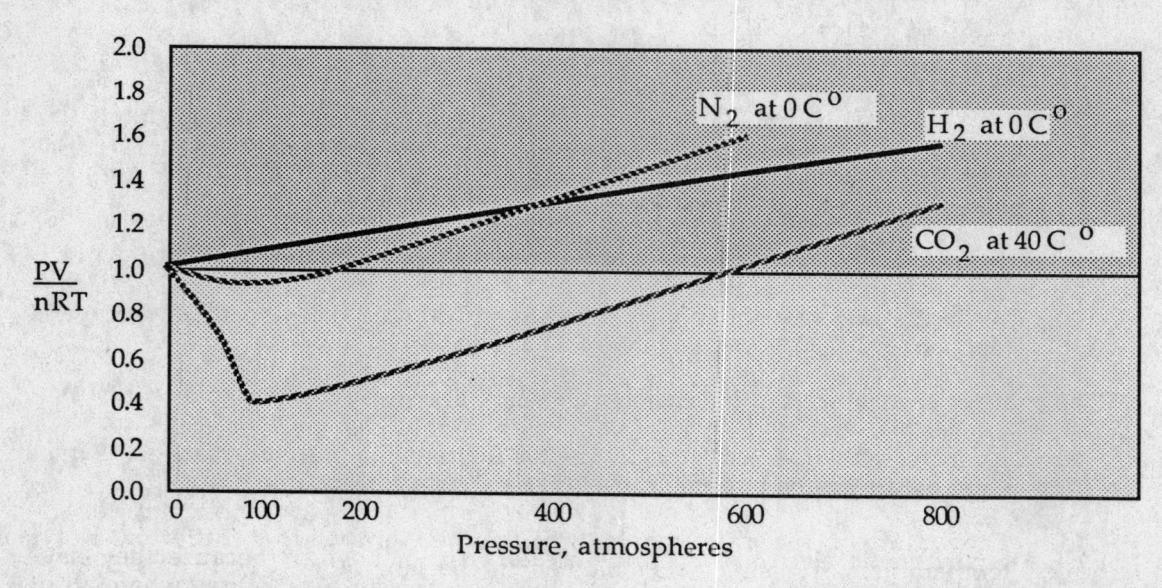

(a) There are two ways that real gases deviate from ideal behavior. List them.

(b) In the temperature range shown (between 0–40°C), compare the molecular attractive forces of nitrogen, hydrogen and carbon dioxide. How does the data plotted in the graph justify the comparison?

(c) At high pressures, the PV/nRT ratio of all three gases exceeds 1.0. Explain this in terms of the kinetic-molecular theory.

(d) At 600 atm, the PV/nRT ratio of nitrogen is greater than that of hydrogen. Does this mean that the attraction between hydrogen molecules is greater than that of hydrogen? Justify your conclusion.

Answers and Explanations

1. (a) The Pauli Exclusion Principle states that:
   "Two, but not more than two electrons can occupy an orbital, and only if they have opposite spin."

   (b) Hund's Rule, restated, is that electrons will half–fill orbitals with parallel spin until all the orbitals are half–filled. Subsequent electrons will pair until the sublevel is filled.

Carbon	2s ⊗	2p	⊗ ◯ ◯	
Nitrogen	2s ⊗	2p	⊗ ⊗ ⊗	
Oxygen	2s ⊗	2p	⊗ ⊗ ⊗	
Neon	2s ⊗	2p	⊗ ⊗ ⊗	

   (c) Paramagnetic electron structures create a magnetic field because they have unpaired electrons. Diamagnetic atoms have electron structures where all of the electrons are paired. Electron pairs cancel the magnetic field.

   (d) The magnetic fields of carbon and nitrogen show paramagnetism because of 2 (carbon) and 3 (nitrogen) unpaired electrons. Neon, with all electrons paired, would be diamagnetic.

2. (a) Electron affinity is a measurement of the amount of energy released as an electron is added to a neutral, isolated gaseous atom.
   The electron affinity increases with increasing atomic number in a Period because electrons are being added to the valence shell at the same time that the number of protons attracting them is increasing.

   (b) The electron affinity of nitrogen is zero because nitrogen has a stable electron structure. The 2s–sublevel is filled and the 2p–sublevel is exactly half–filled. Filled and half–filled electron structures are especially stable.

3. (a)

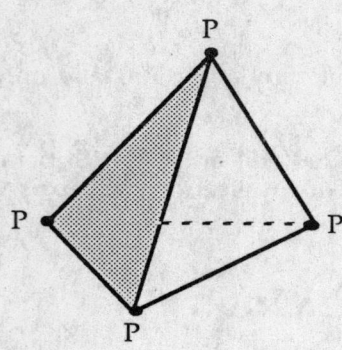

[Tetrahedron with P atoms at each vertex]

(b) Each face of the tetrahedron is an equilateral triangle, so the P—P—P bond angle is 60°.

(c) [Structure of PH₃ showing P with lone pair at top and three H atoms below in trigonal pyramidal arrangement]

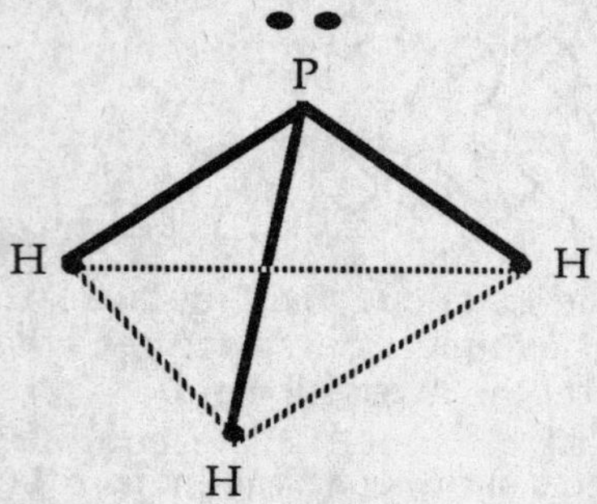

(d) The bond angle is < 109.5°. The H—P—H bond angle is inside a tetrahedron. The bond angle for this location is usually 109.5°, but the lone electron pair will reduce the angle.

4. (a)

cyanide ion $\left[:C \equiv N: \right]^{-}$ nitrite ion $\left[:\ddot{O} — N = \ddot{O}: \right]^{-}$

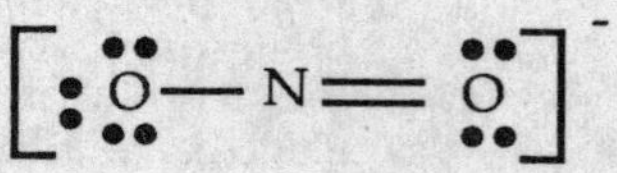

(b) The cyanide ion has one σ bond and two π bonds.
The nitrite ion has two σ bonds and one π bond.

(c) Ligands have lone electron pairs which participate in coordinate covalent bonding with d- orbitals of a metal ion to form complex ions.

(d) Carbon monoxide is structurally similar to the cyanide ion, CN⁻, and would fit into the series of bonding strength of ligands near CN⁻.

carbon monoxide $:C \equiv O:$

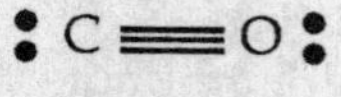

5. (a)
Oxidation: $Tl^+ \Rightarrow Tl^{3+} + 2e^-$
Reduction: $IO_3^- + 2Cl^- + 6H^+ + 4e^- \Rightarrow ICl_2^- + 3H_2O$
Net ionic: $IO_3^- + 2Tl^+ + 2Cl^- + 6H^+ \Leftrightarrow ICl_2^- + 2Tl^{3+} + 3H_2O$

(b) The reducing agent, Tl^+, is oxidized.
The oxidizing agent, IO_3^-, is reduced.

(c) The Nernst equation for this reaction is:

$$\varepsilon = \varepsilon° - \frac{0.059}{4} \log \frac{[ICl_2^-][Tl^{3+}]^2}{[IO_3^-][Tl^+]^2[Cl^-]^2[H^+]^6}$$

This redox reaction is an equilibrium reaction. The reaction is not at equilibrium until the ratio of [Products] to [Reactants] is equal to the equilibrium constant, K.

High values of the net cell potential, ε, occur when the concentration of the products is low and the concentration of the reactants is high. When this occurs the log expression in the Nernst equation will have a negative value and will be added (a negative times a negative) to $\varepsilon°$.

The $[Tl^{3+}]^2$ and $[H^+]^6$ have the highest exponents in the equilibrium law expression, and these should be 0.10 M and 1.50 M respectively.

(d) The free energy change, $\Delta G°$, can be calculated from the net cell potential using the equation:

$$\Delta G° = -nF\varepsilon°$$

The equilibrium constant, K can be calculated from the net cell potential using the equation:

$$\log K = \frac{n}{0.059} \varepsilon°$$

6. (a) $Ag^+(aq) + OH^-(aq) \Rightarrow AgOH(s)$
$Al^{3+}(aq) + 3OH^-(aq) \Rightarrow Al(OH)_3(s)$
$Zn^{2+}(aq) + 2OH^-(aq) \Rightarrow Pb(OH)_2(s)$

(b) Aluminum and zinc hydroxides are amphoteric.
$Al(OH)_3(s) + OH^-(aq) \Rightarrow [Al(OH)_4]^-(aq)$
$Zn(OH)_2(s) + 2OH^-(aq) \Rightarrow [Zn(OH)_4]^{2-}(aq)$

7.

	[C(s)], M	[H$_2$(g)], M	[CH$_4$(g)], M
(a)	Remains the same.	Increases.	Decreases.
(b)	Remains the same.	Decreases, then increases	Decreases.
(c)	Remains the same.	Remains the same.	Remains the same.
(d)	Remains the same.	Increases.	Decreases.
(e)	Remains the same.	Increases.	Increases, then decreases.

Notes:
- (a) An increase in temperature favors the endothermic reaction; in this case the reverse reaction. The concentration of a solid is constant.
- (b) The removal of H$_2$ slows down the forward reaction rate.
- (c) Addition of a solid does not affect the equilibrium.
- (d) Decreasing the pressure favors the reaction that will increase the number of moles of gas.
- (e) The addition of CH$_4$ speeds up the forward reaction rate.

8. (a) $2NaHCO_3(s) \Leftrightarrow Na_2CO_3(s) + CO_2(g) + H_2O(g)$

(b) The sample, by not being heated enough, will not evolve of CO$_2$ and H$_2$O completely. If the amount of gases evolved is too low the computed (from the loss in weight of the sodium bicarbonate) percent NaHCO$_3$ would be low also.

(c) The reported percent sodium bicarbonate, NaHCO$_3$, would be too low.
During the night some CO$_2$ and H$_2$O from the air would be absorbed by the residue, Na$_2$CO$_3$, and its mass would increase. The computed percent NaHCO$_3$ (from the loss in weight) would be too low. The sample should be reheated and cooled just before weighing.

9. (a) Rate = k$_3$[COCl][Cl$_2$]   (from slow step)
       [COCl] = K$_2$[Cl][CO]   (from reaction 2)
       [Cl] = K$_1$$^{1/2}$[Cl$_2$]$^{1/2}$ (from reaction 1)

   Rate = K$_2$k$_3$K$_1$$^{1/2}$[CO][Cl$_2$]$^{3/2}$ = k[CO][Cl$_2$]$^{3/2}$

(b) The units of k must allow the rate units to be mol L^{-1} time^{-1}.
For this reaction, the units of k are L$^{3/2}$ mol$^{-3/2}$ time^{-1}

(c) At the molecular level, when the temperature is increased the average kinetic energy of the molecules increases. This result in more collisions between molecules because their velocity is higher. But, more importantly, there is an increase in the number of molecules with sufficient activation energy to react when a collision occurs.

(d) A form of the Arrhenius equation is:

$$\ln \frac{k_1}{k_2} = - \frac{E_a}{R} \left[\frac{1}{T_2} - \frac{1}{T_1} \right]$$

1. ln k should be the dependent variable (y–axis).
$1/T$ should be the independent variable (x–axis).

2. The slope of the line is $-E_a/R$.

10. (a)

When $\Delta H°$ is positive, the reaction is endothermic.

When $\Delta S°$ is positive, the products are more chaotic.

When $\Delta G°$ is positive, the reaction is nonspontaneous.

(b) The substances in the highest amounts are the reactants, Fe_2O_3 and C. The reaction is not spontantaneous and little, if any, product will be formed.

(c) $\Delta G° = \Delta H° - T\Delta S°$
Increase the temperature of the process. The positive entropy change is favorable for spontaneus reaction, and the $T\Delta S°$ term will be higher at higher temperatures. When $T\Delta S°$ is less than $\Delta H°$ the free energy change, $\Delta G°$, will be negative.

11. (a) 1. Real gas molecules have volume.
Ideal gases molecules take up no space.

2. Real gas molecules attract each other.
Ideal gases have no attractions between molecules.

(b) Carbon dioxide molecules have a higher vanderWaal's attraction than H_2 or N_2. The molecules deflect each other and their paths to the wall are curved rather than straight. As a result they hit the wall less frequently, and the pressure is lower at each temperature.

(c) At higher pressure, the molecular volume becomes more significant. The molecules are closer together and the effective volume (void space) is smaller than expected. There will be a higher collision rate with the walls of the container than for an ideal gas.

(d) The attraction between hydrogen molecules is not greater than that of nitrogen. Hydrogen molecules have fewer electrons, so the molecular attractions between H_2 molecules is lower than those of N_2.
The reason the PV/nRT ratio for nitrogen is higher than hydrogen is that N_2 is a larger molecule.

Appendix

Appendix

A. Summary of Formulas

Chapter 1: Atomic Structure and Periodicity

HALF LIFE (A FIRST ORDER KINETICS REACTION)

$$\ln \frac{N_t}{N_o} = -\lambda t \qquad \lambda = \frac{0.693}{t_{1/2}} \qquad \ln \frac{N_t}{N_o} = -\frac{0.693\, t}{t_{1/2}}$$

N_t is the number of moles at time t.
N_o is the original number of moles.
N_t/N_o is the mole fraction remaining at time t.
λ is the decay constant and $t_{1/2}$ is the half-life.

Chapter 3: Stoichiometry

MOLE LINK

The formula H_2O could be read to mean that there are:

...half as many moles of oxygen as moles of hydrogen.   ($n_{oxygen} = \frac{1}{2}\, n_{hydrogen}$)

...twice as many moles of hydrogen as moles of oxygen.   ($n_{hydrogen} = 2\, n_{oxygen}$)

...three times as many moles of atoms as moles of molecules. ($n_{atoms} = 3\, n_{molecules}$)

SOLUTION CONCENTRATION

Molarity ($\underline{M}$) = $\dfrac{\text{moles solute}\ (n)}{\text{Liter solution}\ (V)}$

Molality (m) = $\dfrac{\text{moles solute}\ (n)}{\text{kilogram solvent (kg)}}$

Mole fraction (x) = $\dfrac{\text{moles solute}}{\text{moles solution}}$

DILUTION

$M_{final} = \dfrac{V_{initial} \times M_{initial}}{V_{final}}$

COLLIGATIVE PROPERTIES

Fp and Bp Change: $\Delta T_f = K_f m$
                    $\Delta T_b = K_b m$

Raoult's Law $P_{solvent} = x_{solvent} P^{\circ}$
               $\Delta P_{solvent} = x_{solute} P^{\circ}$

Chapter 4: States of Matter

GRAHAM'S LAW

$$\frac{\text{Rate of effusion of Gas A}}{\text{Rate of effusion of Gas B}} = \sqrt{\frac{\text{Molecular Weight of Gas B}}{\text{Molecular Weight of Gas A}}}$$

IDEAL GAS LAW; DENSITY OF A GAS; MOLECULAR WEIGHT OF A GAS

$$PV = [n]RT = \left[\frac{g}{MW}\right]RT$$

$$D = \frac{g}{V} = \frac{P * MW}{RT}$$

$$MW = \frac{gRT}{PV}$$

DALTON'S LAW OF PARTIAL PRESSURES

$$P_{Total} = P_{solv1} + P_{solv2}$$

$$P_{Total} = x_{solv1}P°_{solv1} + x_{solv2}P°_{solv2}$$

Chapter 5: Reaction Kinetics

RATE EQUATIONS

$$2A + 4B \Rightarrow 5C + 3D$$

$$\text{Rate} = -\frac{1}{2}\frac{\Delta[A]}{\Delta t} = -\frac{1}{4}\frac{\Delta[B]}{\Delta t} = +\frac{1}{5}\frac{\Delta[C]}{\Delta t} = +\frac{1}{3}\frac{\Delta[D]}{\Delta t}$$

$$3A + B \Rightarrow 2C + D$$

$$\text{Rate} = -\frac{1}{3}\frac{\Delta[A]}{\Delta t} = +\frac{1}{2}\frac{\Delta[C]}{\Delta t} = k[A]^m[B]^n$$

ARRHENIUS EQUATION

$k = Ae^{-E_a/RT}$ where: A is a constant
E_a is the activation energy.

$$\ln\frac{k_1}{k_2} = -\frac{E_a}{R}\left[\frac{1}{T_2} - \frac{1}{T_1}\right]$$

Chapter 6: Equilibrium

EQUATION RELATING K_C AND K_P

$$K_c = K_p \left[\frac{1}{RT}\right]^{\Delta n} \quad \text{where} \quad \Delta n = ((\text{moles of product}) - (\text{moles of reactant}))$$

ARRHENIUS EQUATION

$$\ln \frac{K_1}{K_2} = -\frac{\Delta H}{R}\left[\frac{1}{T_2} - \frac{1}{T_1}\right]$$

IONIZATION CONSTANTS

$K_a \times K_b = K_w = 1.0 \times 10^{-14}$

BUFFERS

$[H^+] = K_a \times \dfrac{[HA]}{[A^-]}$

$pH = pK_a + \log \dfrac{[A^-]}{[HA]}$

NEUTRALIZATION

$M_{base} V_{base} = M_{acid} V_{acid}$

Chapter 7: Thermodynamics

INTERNAL ENERGY CHANGE, HEAT AND WORK

$\Delta E = q + w \qquad q = m_{H_2O} \times \Delta t_{H_2O} \qquad w = P\Delta V = \Delta n_{gas} RT$

ENTHALPY CHANGE

$\Delta H = \Delta E + w = \Delta E + (\Delta n)RT \qquad \Delta H° = \sum \Delta H_f° \text{ (Products)} - \sum \Delta H_f° \text{ (Reactants)}$

ENTROPY AND ENTROPY CHANGE

$S° = \int c_p \, dT = q_p / T \qquad \Delta S° = \sum S°_{products} - \sum S°_{reactants}$

FREE ENERGY CHANGE

$\Delta G° = \Delta H° - \dfrac{T\Delta S°}{1000} \qquad \Delta G° = \sum \Delta G_f° \text{ (Products)} - \sum \Delta G_f° \text{ (Reactants)}$

$\Delta G° = -RT \ln K \qquad \Delta G° = -nF\mathcal{E}°$

Chapter 8: Electrochemistry

NERNST EQUATION

At 25°C: $\qquad \mathcal{E} = \mathcal{E}° - \dfrac{0.0592}{n} \log K$

Data that may be useful in solving the problems.

B.

Standard half-cell potentials at standard conditions. (1.0 M solutions, 1.0 atmosphere pressure and 25°C.)				
Li^+ +	e^-	$\Rightarrow$	Li	-3.04 volts
K^+ +	e^-	$\Rightarrow$	K	-2.92 volts
Na^+ +	e^-	$\Rightarrow$	Na	-2.71 volts
Mg^{2+} +	$2e^-$	$\Rightarrow$	Mg	-2.37 volts
Al^{3+} +	$3e^-$	$\Rightarrow$	Al	-1.66 volts
$2H_2O$ +	$2e^-$	$\Rightarrow$	$H_2 + 2OH^-$	-0.83 volts
Zn^{2+} +	$2e^-$	$\Rightarrow$	Zn	-0.76 volts
Fe^{2+} +	$2e^-$	$\Rightarrow$	Fe	-0.45 volts
Cd^{2+} +	$2e^-$	$\Rightarrow$	Cd	-0.40 volts
Co^{2+} +	$2e^-$	$\Rightarrow$	Co	-0.28 volts
Ni^{2+} +	$2e^-$	$\Rightarrow$	Ni	-0.26 volts
Sn^{2+} +	$2e^-$	$\Rightarrow$	Sn	-0.14 volts
Pb^{2+} +	$2e^-$	$\Rightarrow$	Pb	-0.13 volts
$2H^+$ +	$2e^-$	$\Rightarrow$	H_2	0.00 volts
Cu^{2+} +	$2e^-$	$\Rightarrow$	Cu	0.34 volts
I_2 +	e^-	$\Rightarrow$	$2I^-$	0.54 volts
Hg^{2+} +	$2e^-$	$\Rightarrow$	Hg	0.91 volts
Ag^+ +	e^-	$\Rightarrow$	Ag	0.80 volts
Br_2 +	$2e^-$	$\Rightarrow$	$2Br^-$	1.09 volts
$O_2 + 4H^+$ +	$2e^-$	$\Rightarrow$	$2H_2O$	1.23 volts
Cl_2 +	$2e^-$	$\Rightarrow$	$2Cl^-$	1.36 volts
Au^{3+} +	$3e^-$	$\Rightarrow$	Au	1.50 volts
F_2 +	$2e^-$	$\Rightarrow$	$2F^-$	2.87 volts

C.

Vapor Pressure of Water	
Temperature °C	Vapor Pressure of Water mm Hg
0	4.6
10	9.2
15	12.7
20	17.4
21	18.5
22	19.7
23	20.9
24	22.2
25	23.6
26	25.1
27	26.5
28	28.1
29	29.8
30	31.5
35	41.8
40	55.0
50	92.2
60	149.2
70	233.7
80	355.1
90	525.8
100	760.0

D.

Universal Gas Constant R	8.31 Joules mole^{-1} K^{-1} 62.4 liter mm Hg mole^{-1} K^{-1} 8.31 volt coulomb mole^{-1} K^{-1}	0.0821 Liter atm mol^{-1} K^{-1} 1.99 calories mole^{-1} K^{-1}
1 Faraday	96,500 coulombs	23,060 calories volt^{-1}
1 calorie	4.184 Joules	
1 electron volt atom^{-1}	23.1 kilocalories mole^{-1}	96,500 kiloJoules mole^{-1}
Speed of light (vacuum)	2.998 x 10^8 m sec^{-1}	
ln$_e$ x	= 2.303 log$_{10}$ x	
Planck's constant (h)	= 1.38 x 10^{-23} Joule K^{-1}	
Avogadro's number	= 6.022 x 10^{23} molecules mole^{-1}	
At 25° C	$\frac{RT}{nF} \ln Q$ or $\frac{RT}{nF} \ln K$	= $\frac{0.0591}{n} \log Q$ or $\frac{0.0591}{n} \log K$

Appendix

Periodic Table of the Elements

1 H 1.008																		2 He 4.003
3 Li 6.941	4 Be 9.012											5 B 10.81	6 C 12.01	7 N 14.01	8 O 16.00	9 F 19.00	10 Ne 20.18
11 Na 22.99	12 Mg 24.31											13 Al 26.98	14 Si 28.09	15 P 30.97	16 S 32.06	17 Cl 35.45	18 Ar 39.95
19 K 39.10	20 Ca 40.08	21 Sc 44.96	22 Ti 47.90	23 V 50.94	24 Cr 52.00	25 Mn 54.94	26 Fe 55.85	27 Co 58.93	28 Ni 58.70	29 Cu 63.55	30 Zn 65.38	31 Ga 69.72	32 Ge 72.59	33 As 74.92	34 Se 78.96	35 Br 79.90	36 Kr 83.80
37 Rb 85.47	38 Sr 87.62	39 Y 88.91	40 Zr 91.22	41 Nb 92.91	42 Mo 95.94	43 Tc (98)	44 Ru 101.1	45 Rh 102.9	46 Pd 106.4	47 Ag 107.9	48 Cd 112.4	49 In 114.8	50 Sn 118.7	51 Sb 121.8	52 Te 127.6	53 I 126.9	54 Xe 131.3
55 Cs 132.9	56 Ba 137.3	57 La* 138.9	71 Hf 178.5	73 Ta 180.9	74 W 183.9	75 Re 186.2	76 Os 190.2	77 Ir 192.2	78 Pt 195.1	79 Au 197.0	80 Hg 200.6	81 Tl 204.4	82 Pb 207.2	83 Bi 209.0	84 Po (209)	85 At (210)	86 Rn (222)
87 Fr (223)	88 Ra 226.0	89 Ac** (227)	104 Rf	105 Ha	106 Unh	107 Uns	108	109 Une									

*Lanthanides (Rare Earths)

58 Ce 140.1	59 Pr 140.9	60 Nd 144.2	61 Pm (145)	62 Sm 150.4	63 Eu 152.0	64 Gd 157.3	65 Tb 158.9	66 Dy 162.5	67 Ho 164.9	68 Er 167.3	69 Tm 168.9	70 Yb 173.0	71 Lu 175.0

**Actinides (Transuranium)

90 Th 232.0	91 Pa (231)	92 U 238.0	93 Np (237)	94 Pu (244)	95 Am (243)	96 Cm (247)	97 Bk (247)	98 Cf (251)	99 Es (252)	100 Fm (257)	101 Md (258)	102 No (259)	103 Lr (260)

Alkaline earth metals · Halogens · Transition Metals · Metals · Nonmetals